Whole Life Appraisal for Construction

Roger Flanagan

Carol Jewell

with
George Norman

Blackwell
Science

© 2005 by Blackwell Publishing Ltd

Editorial offices:
Blackwell Science Ltd, 9600 Garsington Road, Oxford OX4 2DQ, UK
   Tel: +44 (0) 1865 776868
Blackwell Publishing Inc., 350 Main Street, Malden, MA 02148-5020, USA
   Tel: +1 781 388 8250
Blackwell Science Asia Pty Ltd, 550 Swanston Street, Carlton, Victoria 3053, Australia
   Tel: +61 (0)3 8359 1011

First published 2005

Library of Congress Cataloging-in-Publication Data
Flanagan, Roger.
   Whole life appraisal in the construction sector/Roger Flanagan, Carol Jewell.– 1st ed.
      p. cm.
   Includes bibliographical references and index.
   ISBN 0-632-05046-2 (pbk.: alk.paper)
   1.   Building – Estimates. I. Jewell, Carol. II. Title.
   TH435.F62 2004
   692–dc22

                                                                      2004009639

ISBN 0-632-05046-2

A catalogue record for this title is available from the British Library

Set in 10 on 13pt Palatino
by Kolam Information Services Pvt. Ltd, Pondicherry, India
Printed and bound in India
by Replika Press Pvt. Ltd, Kundli 131028

The publisher's policy is to use permanent paper from mills that operate a sustainable forestry policy, and which has been manufactured from pulp processed using acid-free and elementary chlorine-free practices. Furthermore, the publisher ensures that the text paper and cover board used have met acceptable environmental accreditation standards.

For further information on Blackwell Publishing, visit our website:
www.thatconstructionsite.com

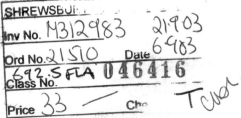

Contents

Acknowledgements

This book has emanted from research undertaken over the past 20 years. We are grateful to organizations who have generously helped and allowed us to use source data. The book is a progression from *Life Cycle Costing for Construction* and we are indebted to the Royal Institution of Chartered Surveyors who allowed us to use some of the ideas from the Flanagan and Norman book.

A special thanks to George Norman, whose original ideas and work have been used. George is based at Tufts University, USA where he is a professor in the Department of Economics. His expertise and knowledge have proved invaluable over the years. The work on discounting is exclusively George's work.

Preface

In the old economy, the pressure was to minimise initial capital cost and, although 'lip service' was paid to the operating and maintenance costs, in effect, they were ignored. In the new economy, everybody is interested in value for money, whether they are the user, investor, developer, design team, construction team, or member of the supply chain. Over a 25-year period an office building will cost about three times its initial capital cost to operate and maintain, yet far more attention is paid to the initial capital outlay than to the significant running costs. It is far too short-sighted to consider just the initial costs when substantial cost and environmental savings can be made over time.

Whole life appraisal is a valuable tool that considers both cost and performance over the whole life of a facility and involves balancing the capital costs against the future operating and maintenance costs whether it is a school, a road, an airport or a power plant. The privatisation of infrastructure and buildings around the world has led to an increase in the number of build-operate-transfer/public private partnerships/concession schemes where a concession is granted to design, finance, build and operate a facility over a defined time horizon of 20–30 years. Balancing whole life cost and performance against the capital costs is crucial to the concession team for the economic viability of the investment.

The whole life concept is neither new nor complicated and there is a general acceptance that it will lead to better decisions on design, refurbishment, and facilities and asset management, and provide whole life economies. However, there remain challenges in the application of whole life techniques due to a number of reasons including the 'data problem', with the paucity of standardised information about cost and performance of facilities in use, the uncertainty of predicting future costs, and performance over a time horizon. The book addresses these challenges and attempts to clarify the techniques and explain the jargon that shrouds the whole life appraisal concept.

Whole life appraisal, when properly understood and used, is a useful and powerful tool. It is a practical tool that enhances, rather than replaces, professional skills. It blends knowledge, judgement, and data to make better, more informed decisions about the future and value for money. The book discusses the benefits and the challenges of using a whole life approach for the appraisal of assets from design through construction and into the operating phase. It explains the principles, techniques and use of whole life appraisal in a straightforward, practical and comprehensible way that will appeal to both students and professionals in the sector.

1 Whole life appraisal – an introduction

1.1 Objective

The objective of this book is to present a discussion on the benefits of using a whole life approach for the appraisal of assets from design through construction and into the operating phase and to present in a straightforward and comprehensible way the principles, techniques and use of whole life appraisal. The intention is:

- ❏ To identify the benefits of a whole life approach to the design team, owners and users of buildings.
- ❏ To explain and illustrate the techniques used in whole life appraisal.
- ❏ To understand the data required for whole life appraisal.
- ❏ To present a set of standardised procedures whereby the whole life approach can be implemented in the construction sector.

1.2 What is whole life appraisal (WLA)?

Whole life appraisal is the systematic consideration of all relevant costs, revenues and performance associated with the acquisition and ownership of an asset over its physical/economic/functional/service/design life. It minimises total expenditure through proper appraisal of costs that will be incurred through the life of the facility.

1.3   Why bother with whole life appraisal?

All clients expect the construction sector to deliver value for money. This means cost-effective solutions, efficient performance and fitness for purpose over the whole life of a facility. The aim for most clients is to ensure the most advantageous combination of capital, maintenance, and operational costs is achieved over the life of the facility.

WLA can be used to:

❏ Compare outline and scheme designs.
❏ Refine the design detailing in the choice of systems and components.
❏ Help management to plan and manage the facility through its planned service life.
❏ Promote realistic budgeting and programming of operation and maintenance activities.

1.4   Is whole life appraisal different to whole life costing (WLC)?

Over time the terminology has changed. Initially, the term was 'costs in use', then 'life cycle costing', which then became 'whole life costing'. Now, the term used globally is 'whole life appraisal' where both costs and performance are considered; it embodies through-life costs. This makes sense as costs are influenced by the performance characteristics of plant and facilities, by the standards of maintenance, and by the periods of occupancy.

1.5   Who uses WLA and WLC?

Everybody involved in owning, developing, or managing a facility should care about balancing initial capital cost with whole life costs. Being realistic, there is always a capital cost restraint – if you have not got the cash today it is pointless saying that, by spending 20% more today, 60% would be saved over the service life of 30 years. But WLA is not about spending more; it is about making the right decision at the outset or even during the operating phase. By spending 20% less, you may still be able to save 60% over the whole life by making informed decisions that analyse all the options.

Property developers who leased property on a full-repairing lease used to focus on capital costs, not running costs, because the tenant took responsibility for the cleaning, maintenance, energy costs, etc. Times have changed – tenants want efficient, reliable and low running costs in facilities that are flexible and easy to adapt. They will balance the rental charge with the whole life costs before signing the lease agreement.

Manufacturers, developers, architects, engineers, contractors, specialist contractors, construction equipment manufacturers and owners all need to think about WLA and WLC.

1.6   When should WLA be undertaken?

Whole life appraisal should be integrated into the design process from concept through to occupancy. As the design process moves through each iteration, ideas are developed and assessed. WLA needs to be embedded in the design process to ensure the best value for money. WLA is a tool, a technique to help make better decisions where options are being considered. It also involves planning to identify which aspects of a facility need to meet performance and maintenance requirements.

Throughout the whole WLA process, communication is essential to ensure that all parties are aware of the assumptions made about the in-use environment, the maintenance requirements and the cost limits.

Start the WLA process early on new build projects. Refurbishment and maintenance of existing facilities require a different approach; start with an estimate of the stock condition.

1.7   What is the link between best value and best price?

Best value should drive every decision. Best price may be attractive in the short run, but it will not necessarily deliver the best facility. Any system that gives an increased service life for lower whole life cost should be welcomed. A higher initial investment may provide greater long-term benefits with reduced maintenance and operating costs.

1.8   Nobody can accurately forecast the future, so how reliable are the results of a whole life appraisal exercise?

Whole life appraisal does not guarantee accurate forecasting. All methods of calculating future costs involve estimates and best guesses; WLA is no different. Be clear about the term 'accurate': accuracy relates to the data available and the ability to measure costs, revenues and performance. Reliability is a better term; it reflects some level of confidence placed on a forecast/estimate. Some information is better than no information about the future; as John Maynard Keynes said 'it is better to be almost right than precisely wrong'.

1.9   How can money spent at different times over the life of a facility be incorporated into WLA?

The discount rate is the main tool for taking tomorrow's money into account. Money in the future is not the same as money today; it has a time value. All future money is discounted (reduced in value) by applying a discount rate selected specifically for the appraisal. Discounted present value or present costs should not be confused with real money.

1.10   What discount rate should be used?

The discount rate has two important functions. First, it enables future costs over a time horizon to be brought to a present value (PV) using standard accounting techniques; in effect it is an exchange rate that converts tomorrow's cost and revenues back to today. Second, by converting costs that occur at both regular and irregular intervals to today, it is possible to compare different options on a comparable basis. Not everybody uses the same discount rate; it varies between organisations. The discount rate is crucial, because if a very high rate is chosen it will swamp all other decisions. The higher the discount rate, the lower the impact of future expenditure. Methodologies for discounting future costs, such as Internal Rate of Return (IRR) and Annual Equivalent Cost (AEC) are discussed later in the book.

1.11   How is inflation taken into account in the calculations?

If future costs in the analysis are expected to rise during the whole life, it is important to include the effects of inflation or escalation in the whole life cost appraisal (see Chapter 5), and to differentiate between a 'nominal' and a 'real' discount rate.

'Nominal' is the discount rate set by the organisation; it may be the bank overdraft rate, the bank interest rate for borrowing, or the weighted average cost of capital. The 'real' rate takes general inflation into account by adjusting the discount rate (this is discussed later).

A number of approaches are possible:

❑ General inflation may be considered in setting the discount rate, thus producing a 'real' rate that takes into account both inflation and discounting of future costs.
❑ If a 'real' discount rate is not used, each of the costs over the life in the analysis should be escalated as necessary before the discount rate is applied to them.
❑ Even where a 'real' discount rate is used, it may be necessary to apply differential cost escalation factors to costs that are rising (or falling) faster than the general inflation rate. A common example is the costs of energy   even if say a 4% discount rate is chosen to account for the general effects of inflation, it may be necessary to think of energy costs as escalating 3%, or 5%, or 10% faster than the general inflation rate.

However it is done, it is important to make careful assumptions about inflation and rising costs. If a real rate is used (say, inflation of 3% per year) then its effect becomes greatly magnified if the analysis covers a number of years. Since future inflation and escalation rates are, by definition, estimates, it is important to carry out some 'sensitivity' testing to make sure that assumptions about inflation are not skewing the results of the analysis.

1.12   Why isn't WLA used more widely?

There are a number of reasons for WLA not being used widely, namely:

- ❑ Availability of reliable data on costs and performance, that is relevant, reliable, up to date, cost effective, and can be tested.
- ❑ Lack of awareness of the importance of future costs/values.
- ❑ Concern over the uncertainty surrounding forecasting future events over the life of a facility and its components.
- ❑ The tradition of separating construction and maintenance costs.
- ❑ The complex and theoretical relationship between money now and money spent or received in the future.
- ❑ The diverse nature of the industry's clients, with very different motivation.
- ❑ The long time lag between the design process and data becoming available on the running costs of the project in use, with the data structured in a useful format.
- ❑ The natural concentration by construction consultants on services for which they are paid and, therefore, those in demand.

1.13   So why bother now with WLA?

In the UK, the Private Finance Initiative (PFI) means that some projects are procured on a design, build, finance, and operate basis over a concession period of, say, 25 years. The company formed to build and operate the facility is responsible for the

> The aim of PFI is to secure private sector funding for major building projects, such as schools, hospitals and roads, which would traditionally have been funded by the public sector, either through tax and other state revenue and/or public borrowing. Under PFI, a private sector consortium – normally made up of a construction company, a bank or financier, a facilities management contractor and consultants – finances the capital costs, carries out the construction of the project, provides for its long-term maintenance and manages the support services, in return for a stream of income over a number of years related to the use made of the project. In order for the scheme to be attractive to the private sector, a satisfactory level of profit is guaranteed. In the case of a school being built by PFI, the government pays the consortium a regular fee for the use of the school, which covers the construction costs, the rent of the building, the costs of support services and a profit element to offset the risks transferred to the private sector.

operating costs over the concession period. Therefore, whole life costs and performance are crucial.

Similarly, the various forms of build, operate and transfer (BOT) schemes throughout the world also give the responsibility of operating a facility to the team that designed, financed and built it. Whole life considerations have become crucial, not only in economic terms, but also because of environmental issues and the concepts of sustainability.

1.14   Has the computer made a difference to the calculation of WLA?

Computers have transformed the management and manipulation of data. Databases, statistical packages, search engines, spreadsheets and simulation have all made a difference to WLA. The internet and web portals mean that data can be moved in real time. Manufacturers' data on performance can be downloaded electronically on to a computer and used in an appraisal. The cost, speed, and availability of computing power enhances the use of WLA.

1.15   What has happened overseas?

- ❏ Sweden has extensively used WLA for public sector projects; it has a whole life culture.
- ❏ Japan has a WLA policy for public sector expenditure.
- ❏ Australia requires public sector projects to have a whole life appraisal.
- ❏ Where the USA has influence, whole life is on the agenda. Some South American countries are looking into whole life issues.
- ❏ Developing countries are so short of cash that whole life considerations are not part of the equation. However, on internationally funded projects such countries are forced to justify the project on a whole life basis. The World Bank requires all investment appraisals to include a WLA before the project is financed.
- ❏ There has been a move to provide guarantees on projects, beyond the defects liability period. Selling guarantees is

becoming good business, but it requires an understanding of whole life issues.
- ❑ Facilities and asset management is a growing industry with a much better view on whole life costs and performance.
- ❑ BOT is growing in use to procure many types of projects, hence WLA is an important part of balancing the initial capital cost with the whole life cost.

1.16   The whole process seems to be surrounded by jargon; how can WLA be made simpler to understand?

Common sense must prevail. Discounting can be confusing; however, the underlying concept is very straightforward: money has a time value. Take the approach on a step-by-step basis. Running costs is the term used to describe all the operating and maintenance costs of an asset over its life.

1.17   Where does sustainability fit with WLA?

Business and projects must be sustainable. Twenty years ago there was little awareness of the importance of sustainability; now it is high on everyone's agenda. Sustainability means thinking about how a project affects the environment, now and in the future. There needs to be an increased use of materials that can be reused and recycled, a reduction in waste and emissions and more use of renewable energy alongside energy efficiency measures. The effect of sustainability on the whole life of a project means that it is best integrated into the design process. However, this does not preclude making decisions based on sustainability in the repair, maintenance and refurbishment processes.

1.18   Have the Construction Design and Management Regulations (CDM) 1994 changed the way whole life issues are considered?

The CDM Regulations (HMSO, 1994) in the UK place a duty on clients, designers and construction personnel to consider health and safety hazards during construction and maintenance. Regulation 13(2)(a)(ii) states that every designer should ensure that any design takes adequate regard to combat, at source, risks to the health and safety of any person carrying out construction work or cleaning work in or on the structure at any time. This means that maintenance and cleaning must be considered at the outset to ensure workers are not put at risk. Hence, both capital and ongoing costs need to be considered.

1.19   How does WLA fit with facilities/asset management?

Facilities and asset management is concerned with ensuring a facility/asset has fitness for purpose, is maintained effectively, and provides value for money. WLA is a tool to be used by the facility/asset management team to ensure value for money is provided. ISO 14040 provides a methodology for life cycle assessment.

1.20   What is service life planning?

Service life planning is an integral part of WLA. It involves consideration of the likely performance of a facility over the whole of its life or the chosen time horizon under the applicable environmental conditions, from concept through to occupancy. If the service life is less than the design life (the service life chosen by the designer) then it should be replaceable or maintainable. ISO (the International Organisation for Standardisation) 15686 provides a methodology for forecasting the service life and estimating the maintenance and replacement requirements. Performance and safety are clearly linked; priority should be given to ensuring performance at all times meets national legislative requirements for safety.

1.21   Integrated Logistic Support (ILS) is a term used by the defence and aerospace industries; is WLA used in ILS?

Whole life appraisal is an integral part of ILS, which is described in detail in a later chapter; basically, it involves managing facilities/equipment such as a tank or an aircraft over its life. The tank is a complex piece of equipment that operates in hostile environments; it has to be maintained and, should anything malfunction, the team need to know the availability of spare parts and even the size of spanner needed to undo the nuts. WLA is used in ILS as a tool.

1.22 What is the difference between running costs and operating and maintenance costs?

None. The terms are used interchangeably to reflect the cost of owning and operating a facility/asset. The term operating and maintenance costs has been used in this book to standardise terminology.

1.23 How does the residual value work?

Every facility, component, project will have a residual value, it may only be the scrap value of the material, or it may be the site value, that has increased significantly. Residual value is a 'best guess' about what will be a value at the end of a period of analysis. A boiler that has given 20 years' service may not even have a scrap value and will attract a cost for its removal. An office building may have increased in value, but after 30 years, the site value may be worth more for an alternative use.

1.24 Should disruption costs be included in the WLA?

Whole life costs are both tangible and intangible. In some situations, unplanned maintenance work can cause disruption costs far in excess of the cost of maintenance. In hospitals, airports and ports, the magnitude of the disruption costs will far outweigh the maintenance costs. Where appropriate, allowance for disruption costs should be included in the calculations to fully reflect the whole life cost.

2 Whole life appraisal: preliminary concepts

2.1 Introduction

Whole life appraisal involves the consideration of the costs, benefits, and performance of a facility/asset over its life. The building industry tends to use the term facility to represent buildings and properties, whereas the construction industry which includes civil engineering, tends to use the term asset management to include roads, rail, ports and harbours, and coastal engineering. For simplicity, facility/asset management is used throughout this book.

There are a number of definitions of whole life costing:

❑ The total cost of a facility/asset over its operating life including initial acquisition costs and subsequent running costs.

❑ The systematic consideration of all relevant costs and revenues associated with the acquisition and ownership of an asset (Construction Best Practice Programme, 2001; CRISP Performance Theme Group, 1999).

❑ The sum of present values of investment costs, capital costs, installation costs, energy costs, operating costs, maintenance costs, and disposal costs over the life time of the project, product, or measure.

❑ A tool to assist in assessing the cost performance of construction work, aimed at facilitating choices where there are alternative means of achieving the client's objectives and where those alternatives differ, not only in their initial costs but also in their subsequent operational costs (ISO 15686-1, 2000)

Over a 25-year period an office building will consume about three times its initial capital cost, yet far more attention is paid to the initial capital outlay than to the significant running costs. It is far too shortsighted to consider just the initial costs when substantial cost and environmental savings can be made over time. A Royal Academy of Engineering report found that the typical costs for owning a building were in the ratio of, 1 for construction costs : 5 for maintenance and building operating costs : 200 for business operating costs (Royal Academy of Engineering, 1998).

Value is about considering the 1 : 5 : 200 rule where:
    1 = the initial cost of a facility
    5 = operating costs/cost in use over 20 years
    200 = the value of the business done in the facility over 20
        years
Conventional commercial development concentrates on 1 and ignores the rest. PFI is based on 5.

A good example is a patient in a hospital. If they have pleasant surroundings in a good environment with a well-designed modern building, they may recover more quickly and be discharged from hospital sooner. Everybody benefits as the hospital will be able to use the bed for another patient, thus increasing efficiency, and the patient may get back to work earlier.

Whole life appraisal is not new; it has used tried and tested techniques applied to spheres of many industries over a number of years. Even at the simplest level, whole life costs are taken into consideration – for instance, nobody would purchase a car that they could not afford to operate and maintain. They would be concerned about the fuel consumption, the servicing, maintenance and repair costs, the insurance cost, and the residual value when the car was sold. The retail and leisure industries have adopted a whole life costing with its dependence on repair and maintenance to maintain an attractive and dynamic image. Large construction clients/donors such as the World Bank and the European Bank for Reconstruction and Development (EBRD) insist on whole life appraisal for proposed projects.

A whole life and performance approach that takes explicit account of the whole life cost and performance of assets, is essential to effective decision making in four main ways.

If a building, costing €5 million is in use for 15 years, with an operating cost of €500 000 p.a., it starts to consume money from day one. Over its service life, it will cost around €15 million, allowing for inflation and all the running costs. If only 5% could be saved on operating and maintenance costs, the saving would amount to €750 000.

❏ It identifies the total cost commitment undertaken in the acquisition of any asset, rather than merely concentrating on the initial capital costs.

❏ It facilitates an effective choice between alternative methods of achieving a stated objective, e.g. choosing between two different machines to perform a particular production process, or two different designs for a commercial building. It takes full account of the probability that the various options are likely to exhibit somewhat different patterns of capital and running costs, and provides a set of techniques for expressing those costs in consistent, comparable terms.

❏ A whole life approach is a management tool that details the current operating costs of assets such as machinery, individual elements (heating systems, roof coverings), or complete systems such as road surfacing.

❏ Whole life appraisal identifies those areas in which operating costs might be reduced, either by a change in operating practice, for example hours of operation, or by changing the relevant system.

Whole life appraisal can be used to draw up a maintenance programme and to identify the sinking fund to finance a planned maintenance programme.

While whole life appraisal techniques can be applied in any area of economic decision making, they are particularly relevant to the proper identification and evaluation of the costs of durable assets. As a result, they are of especial relevance to the construction sector. Whether complete facilities or individual elements, or infrastructure projects are considered, a decision is being made to acquire assets that are intended to last and to be used for many years. These assets will commit the owner or user not only to initial capital costs, but also to subsequent running costs, day-to-day operating, cleaning and maintenance costs, periodic repair or replacement costs, energy costs, water and sewerage costs, insurances, and local taxes. Equally importantly, decisions made at the initial design stage will invariably affect future running costs and the economic use of the facility. For instance, there are numerous ways to heat a building, to illuminate it, to clad it, and divide the space into workable areas, each with different initial and running cost profiles.

> Contractors involved in PFI projects are more aware of whole life costs as they have to calculate the service life of a facility in their bid, then they have to operate the facility over a concession period (often 20–25 years).

New methods of procurement and the growing concerns of informed owners and operators about the costs of their facilities have focused attention on whole life. Owners and users should, therefore, be encouraged to look upon their assets in the same way as any other productive units such as plant, machinery or equipment. The cost of ownership should be planned and managed throughout the life cycle, but especially at the early design stage.

The basic premise of whole life cost is that, to an investor, all costs (future as well as present) arising from an investment decision are potentially important. Thus whole life appraisal should be seen as essential in the decision-making process from the beginning, and must be considered as a major evaluation criterion in the design brief.

One major complication arises in the adoption of such a total whole life approach. The relevant costs are a combination of capital costs (incurred at the initial acquisition stage) and running costs (incurred at varying points during the subsequent operation of the building or building component). Since these costs are incurred at different times they cannot be treated identically, 'money today' not being equivalent to 'money tomorrow'. A whole life approach must have as a central feature the presentation of current and future costs in equivalent terms.

> ISO 15663-1 (2000) states a number of benefits in applying the systematic application of whole life costing:
>
> ❏ Reduced ownership costs.
> ❏ The alignment of engineering decisions with corporate and business objectives.

- ❏ The definition of common objective criteria that can be used by operators, contractors and vendors against which, business transactions may be managed and optimised.
- ❏ Reduction of the risk of operating expenditure surprise.
- ❏ Changing the criteria for option selection.
- ❏ Maximisation of the value of current operating experience.
- ❏ The provision of a framework within which to compare options at all stages of development.
- ❏ The provision of a mechanism by which major cost drivers can be identified, targeted and reduced.

2.2   Whole life appraisal and construction

This decision method has the undoubted merit of (relative) simplicity, and has been justified on two main grounds:

- ❏ Initial capital cost is the single, most important cost commitment undertaken when purchasing facilities; all other costs are 'unimportant' and can be ignored.
- ❏ Since capital cost is the single most important cost, the lowest capital cost option will also be the lowest total cost option; the implication is that there are no real benefits to be gained from reducing running costs by increasing capital costs.

Both arguments are now being challenged. Even if capital costs were dominant in the past (and this is open to question) it is being forcefully argued that this is no longer the case. Two stories illustrate this point:

- ❏ A lead story from a provincial newspaper stated that the 'Local municipality's running cost for a school in a rural area with a declining number of pupils reached €500 000 last year'. The question is being posed as to how much of the €500 000 could have been reduced or avoided by a total cost approach at the design stage, recognising there was a decline in pupil numbers.
- ❏ A major industrial client stated that if the running costs of the warehouse had been calculated at the design stage, a completely different building would have been commissioned.

The point of both stories is that, while initial costs are clear and visible at an early stage, longer-term costs are not – see Figure 2.1. Nevertheless, these longer-term costs can far outweigh initial capital costs, and should have a much stronger influence on decisions with respect to facilities and individual elements.

Figure 2.2 illustrates the whole life cost commitment for a small infants school built in the UK. In this figure, running costs, which are incurred either annually or at periodic intervals over the life of the school, are compared with capital costs, which are incurred at the initial construction stage. As was noted above, for such a comparison to be valid, future costs have to be converted (discounted) to their current equivalent. The discounting techniques for making this conversion are presented in Chapter 5. Suffice it to say at this stage the owner of the school applied a 4% real discount rate (net of inflation) to all future costs. The time horizon for the appraisal of the school – the whole life of the analysis – has been assumed to be 30 years. It shows that capital costs account for well under half of the total cost commitment.

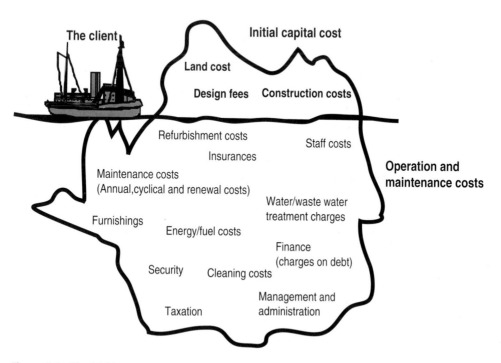

Figure 2.1 The hidden costs.

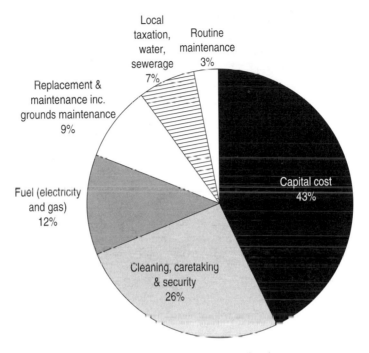

Figure 2.2 The whole life costs of a primary school.

Why was 30 years chosen for the analysis? This is the owner's choice; much of the physical structure would last between 50 and 100 years with proper maintenance. The horizon of 30 years may reflect the owner's view of how long the facility can be used without major retrofitting and change. Alternatively, it may reflect the length of time the owner has used for the investment appraisal of the school.

Figure 2.3 illustrates the whole life cost commitment for a new flooring system installed in an office facility. In this case a 4% real discount rate and a period of analysis of 25 years has been used. Figure 2.3 shows that concentration solely on initial capital cost will give a very imperfect view of the actual costs being incurred.

2.3   Performance of systems, components and materials

Cost and performance of the facility are linked. The fuel cost for the heating and cooling is affected by the pattern of use (hours the facility is in use, the operating temperature selected), the level of maintenance (a boiler that has not been serviced

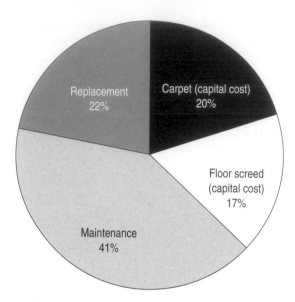

Figure 2.3   The whole life costs of a floor covering.

regularly will not operate at maximum efficiency), the maximum and minimum daily outside temperature, the aspect of
the facility (north facing rooms will need more heat), and the
level of insulation.

Information/data is needed from manufacturers, and installers
on performance. Feedback will be discussed later. The challenge
is to ensure the feedback data are relevant and appropriate.

2.4   Design with whole life appraisal

Whole life techniques are used as a tool in the design of facilities.
Some designers make whole life cost allowances in particular
types of construction. Nevertheless, it is fair to argue that the use
of the techniques remains unsatisfactory. Two reasons for this are:

❏   The lack of familiarity with whole life appraisal.
❏   Costs have traditionally been separated into boxes of capital,
     maintenance, operating, and disruption costs.

> Some 90% of the cost of running, maintaining and repairing a
> facility is determined at the design stage.

In addition, there are important institutional constraints that must be overcome. Many public sector clients, for example, are required to use capital cost as the sole criterion in decision making. Indeed, most public sector funding arrangements separate capital and revenue budgets, thus imposing severe restrictions on a total cost approach; however, this is changing and Treasury guidelines support a whole life approach to investment.

The argument that the lowest capital cost option will also be the lowest total cost option is open to question. It is based in part on the relative dominance of capital costs, and the discussion above indicates how weak that argument is. Any number of recent examples can be produced to challenge the 'minimum capital cost' approach to decision making.

In other areas it can also be shown that a total cost approach can generate significant cost savings. An example is outlined in Figure 2.4. Two floor-covering options were presented to a client – a wood-block floor or carpet over a screeded floor. Over 30 years the wood-block covering is preferred in a total cost approach, despite this option having the higher initial cost. It offers significant savings on annual maintenance costs – particularly on cleaning costs – that are sufficient to offset the additional initial and subsequent replacement costs.

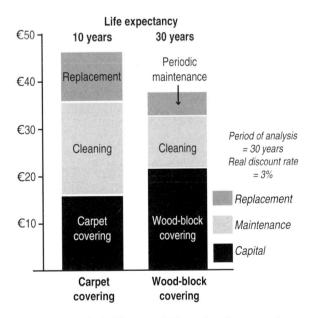

Figure 2.4 Whole life cost of alternative floor coverings.

Fashion and flexibility cannot be ignored. Fashion changes and some clients may prefer the flexibility of having a covering that is quickly and easily changed.

2.5   Whole life appraisal for new and old facilties

It might be thought from this discussion that the main area of application for whole life appraisal is with respect to new buildings and infrastructure, or the individual elements and components. This is not the case. A whole life approach also offers potential cost savings when applied to existing assets. Indeed, this is probably one of the most important areas of application of whole life techniques in the construction sector.

Buildings and infrastructure are durable assets, and while new facilities are continually being added to the existing stock, the majority of facility users will be making decisions with regard to existing facilities. At the same time, the durability of buildings implies that design decisions made during their initial construction, while perhaps correct at the time, may well need to be altered to adjust for unforeseen changes in economic conditions. In an era of low energy costs it would have been difficult to justify double or triple glazing solely on cost grounds, whereas it is now sensible to change existing glazing systems. The move towards materials that are easy to clean and maintain is driven at least in part by the increase in labour costs relative to other costs. A whole life approach identifies the potential for such design changes.

Systematic analysis is required; let us consider windows as an example:

- ❑ Timber windows with proper maintenance will last between 30 and 40 years, often longer.
- ❑ Metal windows will last 30 years or longer.
- ❑ Plastic PVC windows will last at least 30 years.
- ❑ Double-glazed units fitted into windows will last between 10 and 20 years; it is the edge seals that fail, not the glass.
- ❑ Energy-saving glass in double-glazed units will reduce heat loss and increase the solar heat gain; in double-glazed units, the seals will last between 10 and 15 years.
- ❑ Self-cleansing glass is new, so it is not known how long the coating will last.

- ❏ Mastic sealant around the window frame will need replacing every 10 years.
- ❏ Fashion will dictate that dated designs may be replaced after 20 years with a more modern window design.

Hence, over a 30-year period of analysis, it is the seals on double glazing that are likely to fail. Furthermore, windows on an exposed south face in a building are likely to fail faster than those on the cooler north face. Judgement, skill and knowledge are required to ensure realistic forecasts are made.

For a whole life approach to be effective in reducing the running costs of existing facilities it is necessary that these running costs are monitored. In other words, whole life cost should be seen also as an essential element of overall cost management. It can identify in detail one major area in which an organisation incurs costs – the operation of assets – and point to ways in which potential cost savings might be achieved.

While the 'minimum capital cost' approach might not have been totally wrong in the past, this can no longer be assumed to be the case. It is of increasing importance that a whole life approach be adopted at the early design stage and in the management of the existing stock.

Whole life cost will loosen the decision maker's traditional concentration on capital costs to the almost total exclusion of any concern with future running costs. There must be a recognition that decisions made today carry cost implications in the future, whether these decisions relate to the design of a complete building, or to the choice of an individual component. Further, once an organisation begins to apply whole life techniques it will develop an historical database that will make it easier to identify future areas of potential cost reduction.

This is not to say that the implementation of whole life appraisal in the construction sector will be straightforward. The nature of the construction process and the length of time separating the design and the operation/user phases make the transmission of user performance data more difficult than in other industries. In addition buildings and infrastructure are complex, and individual building elements interact in a diversity of ways, for example, a change in plan shape or building aspect will affect not only the initial construction costs, but also subsequent heating, lighting and other operating costs.

3 An overview of whole life appraisal

3.1 The importance of whole life appraisal

Much of the industry deals with the project process as a series of sequential and largely separate operations undertaken by individual designers, constructors and suppliers who have little or no involvement in the long-term operation of the product and in its operating and maintenance costs. Figure 3.1 shows the range of whole life costs involved in the construction and operation of a facility.

Owners/clients want to know more about the balance between the initial capital cost and the whole life cost and performance with traditional design–bid–build. New procurement routes are being adopted where the designer is part of a consortium that is carrying design, construction, and operating and maintenance risk over the whole life of a project. Examples of these are public private partnerships, private finance initiative (PFI), and build, operate and transfer (BOT). Similarly, on design–build projects in a variety of guises, there is a single point responsibility for the delivery of the project where the owner will seek certain performance requirements of the project in use.

3.2 The adoption of whole life appraisal

It may seem obvious to suggest that the decision maker should examine the total cost implications of any decision. After all, consumers apply this concept in a more or less formal fashion when considering expenditure on consumer durables such as

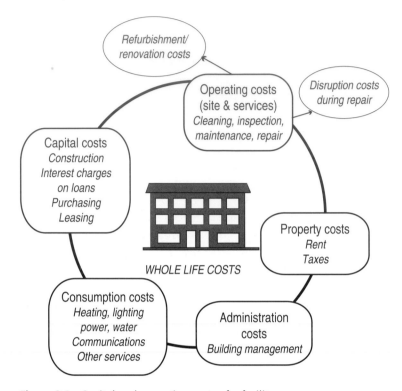

Figure 3.1 Capital and operating costs of a facility.

cars, freezers or houses. The consequences of not doing so are clear. Nevertheless, it is probably true to state that within the UK, whole life cost has not been generally implemented within the construction sector. Several reasons for this apparent failing can be suggested as shown under the following headings.

Awareness

Awareness on the part of the client cannot be discounted. The majority of clients are not 'in the business' of construction; they require a facility, whether it is a factory, office or commercial premises, road or runway as an input to their main productive activity. The onus lies increasingly with the clients' advisers to ensure that clients are made fully aware of the options open to them, ranging from demolition of an existing facility and replacing with new, partial demolition and partial refurbishment, through to retaining and refurbishing; or even selling the facility and finding a new site.

Balance between capital and operations

The main impetus to whole life appraisal in construction emanates from the increase in relative importance of operating and maintenance costs and the performance of the facility. Changes in the economic climate have upset the balance between capital and operating and maintenance costs of facilities. It is reasonable to expect that real incomes, fuel costs, insurances, security costs, utility costs and labour costs will continue to rise, further increasing the importance of operating and maintenance costs relative to capital costs. Insurance costs have risen dramatically since the September 11th, 2001 terrorist attacks, with annual premiums for some types of insurance rising more than 100% over a two-year period.

Separation of capital and operating budgets into different boxes

The classic form of whole life appraisal cuts across the traditional separation of capital and operating budget functions. Capital and operating costs are linked and should not be treated separately. It follows that if management decisions and control systems are to be compatible with whole life output and recommendations, there must be some integration of capital and operating budget procedures.

Collecting data on cost and performance

Standard methods for collecting relevant data are used, to determine just what are 'relevant data', and to develop standard techniques and documentation by which these data can be analysed, updated and used.

Even where data are available, additional efforts are required by the principal decision makers in the design team to analyse the user performance costs. The problem is not so much a lack of ideas, ideas are historically what design teams are best at providing. The task confronting them is dealing with limited economic resources in such a way that optimum proposals are selected and implemented.

The interaction between the operating categories

The output of the construction sector is complex, often sophisticated, and generally non-standardised. Decisions with respect to one element carry implications, both directly and indirectly, for many other elements. Similarly, for an infrastructure project such as a road, bridge, or wastewater treatment plant, many items will be interrelated. These interactions and interrelationships are often difficult to untangle but dangerous to ignore, as indicated by the schematic diagram, Figure 3.2, which relates to the operating costs of a facility.

Thus, there are not only connections between initial capital costs and subsequent running costs, but also between the many elements that make up recurrent operations and maintenance costs. For example, more frequent cleaning of light sources will increase luminosity and so reduce costs. At the same time, different decisions with respect to lighting will affect future cleaning costs, and maintenance costs.

3.3   The balance between fixed and variable costs

The relative balance between fixed (initial capital) costs and variable (operations and maintenance) costs has changed and there is no reason to believe that labour, materials and energy costs will fall in relative terms in the future. Figure 3.3(a) shows alternative design options where the efficient design options were to be ranked in ascending order of their capital costs, and so in descending order of their recurrent operating and maintenance costs (all expressed in current values). Figure 3.3(a) might have been characteristic of cost conditions in the past: the minimum capital cost design option also being the minimum total (whole life) cost option. But the more likely case in future is that illustrated in Figure 3.3(b). It will often, perhaps typically, be preferable to incur higher initial capital costs in order to secure lower future running costs.

If whole life appraisal is to be effective it must be implemented as early as possible in the design process. As can be

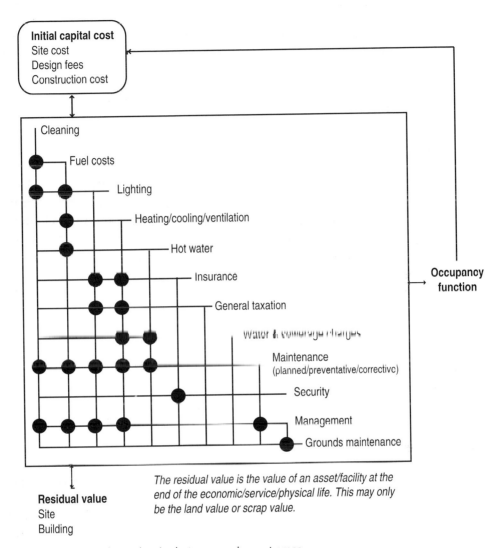

Figure 3.2 Interdependencies between major cost areas.

seen from Figure 3.4, the later the whole life cost techniques are introduced during the design process the lower will be the potential for cost savings and the more expensive it will be to implement any design changes suggested by the results of the analysis.

Two words of caution, first, it might be felt that whole life techniques will automatically reject the least capital cost option: this is wrong. There will continue to be many cases in which such an option is also the least total cost option. Whole life

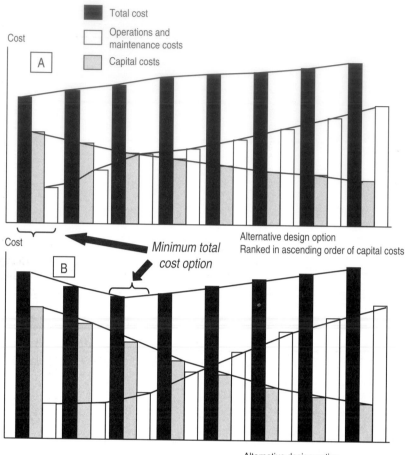

Figure 3.3   The changing balance between fixed and variable costs for various options.

appraisal ensures that selection is made for the correct reasons after proper comparison.

Second, even everyday economic decisions such as the choice between double or triple glazing, or glazing with a specially coated or manufactured glass, may often have to be made by intuitive judgement.

Any estimate about the future costs involves uncertainty and this will be discussed later. A view has to be taken about future costs and performance, even these are uncertain, since they are an essential part of whole life appraisal. Forecasting the future is shrouded in uncertainty, but any forecast is better than nothing.

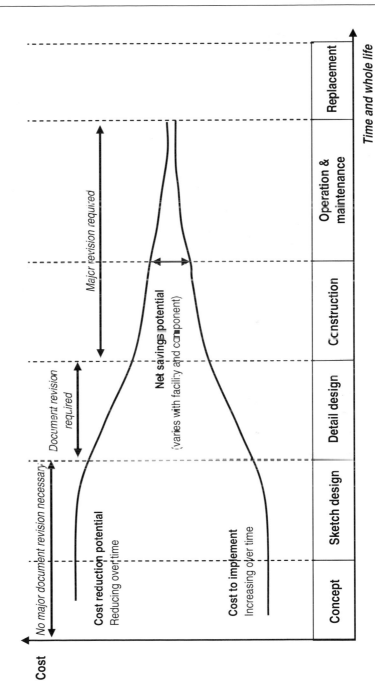

Figure 3.4   Relationship between life cycle cost savings and the timing of implementation.

3.4   Different client motivations for the use of whole life appraisal

> Whole life appraisal is not the universal panacea for the construction sector, but when properly understood and used it is a useful and powerful tool.

Clients of the construction sector cover a wide range of interests and have different requirements and motivations. The reasons for using whole life vary with the type of client. For example, a public sector client usually builds to satisfy a public need whereas the private sector client will be much more influenced by the impact of the project on profitability.

In general terms clients can be classified as follows:

- ❑ Private sector owner-occupiers who commission projects for their own occupation.
- ❑ Public sector owner-occupiers who commission projects for their own occupation. This group will include central and local government and publicly funded organisations such as universities.
- ❑ Public sector clients who build for sale, lease and rental, such as housing associations.
- ❑ Public sector clients who construct for public use such as roads, harbours and ports.
- ❑ Private sector utility companies who construct infrastructure projects such as water treatment plants, railways and communication installations.
- ❑ Developers whose prime function is to develop and finance buildings for sale or for rent, either speculatively or for a known client. There has been a growing tendency for developers to provide a sophisticated service in assembling the ingredients for a development, undertaking it, and subsequently managing it in return for a share of occupational rents from the institution funding the development, or alternatively for a fee based on performance.
- ❑ Financial institutions that include pension funds, insurance companies, property unit trusts, charities and banks, which provide finance for the development projects. Some financial institutions have become directly involved in development.

However, institutions will take strictly limited risks, prefer-ably freehold or with a lease. The institutions will also often purchase completed buildings as investments.

❑ Lessee-occupier who rents space in a building on the basis of a landlord and tenant relationship. This system allows each party to own a separate legal interest that can be managed or marketed unilaterally.

❑ Special purpose vehicle (SPV) company formed to undertake a PFI or BOT project for a road, prison, school, hospital, bridge, tunnel or airport.

The people/organisations – all with different objectives

❑ Investor – Financier
❑ Developer
❑ Owner
❑ Design team
❑ Contractors
❑ Specialist contractor
❑ Maintenance staff
❑ User

The different experience of particular types of client will also affect their perception of whole life, and implies a further classi-fication of clients into those who:

❑ Build only once in their lives.
❑ Have a limited development programme and build once every two or three years.
❑ Have a large continuing development programme.

The whole life process must take into account not only the client type, but also the investor motivation because some clients may choose to invest irrespective of the total cost consequences. Industrialists invest in buildings primarily to meet additional demand or as a consequence of technological change. The pur-chase of new machinery may dictate the need for additional floorspace or alterations to existing buildings to allow for more height, air conditioning, and so on. The owner-occupier has similar needs. For example, the retail store owner might invest to increase the sales area or to create a new image for the company. The question of the client image may be important as many clients are prepared to pay for prestige value.

Public sector clients usually build to satisfy a public need, for example, for housing, sports complexes, schools, fire stations, roads, prisons. Developers and financial institutions, on the other hand, invest primarily for the capital appreciation potential and the income cash flow. In addition, certain types of industrial buildings offer tax advantages to the owner (this is discussed in Chapter 9).

Given this diversity of interests, the incentive for each type of client to apply whole life techniques is understandably different.

Owner-occupier

For the owner-occupier the incentive to apply whole life techniques is obvious. This type of client should treat a building and individual building elements precisely as they would treat any other factor of production in their main activity, whether for the public or private sector.

Lessee-occupier

The lessee-occupier may appear to have little obvious incentive to apply whole life to a complete building, but they should apply whole life to individual decisions in the use of the building systems, for example choice of floor finishings or installation of energy control systems. In addition, if the lessee-occupier is educated to think in whole life terms they will be more attracted to efficient buildings and building systems. Such buildings will offer savings on annual service charges and general running costs, savings that will subsequently be reflected in the rents these buildings command.

Developers or financial institutions

When developers or financial institutions commission buildings for their own use, the incentive to use whole life appraisal is that which applies to any owner-occupier. The relevance of whole

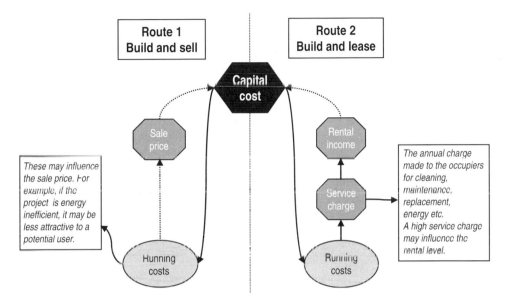

Figure 3.5 The developer's decision

life costs to such clients when they are developing to lease or sell is less obvious, but nonetheless strong. The developer, for example, can take either of the routes illustrated in Figure 3.5. If the developer is commissioning a building for sale to an owner-occupier, the latter will be interested in the whole life of the building. This will feed back, therefore, into the purchase price they are willing to pay. The developer should be able, by appropriate marketing, to reap the advantages of commissioning an efficient building.

A similar argument applies to the case in which the developer rents the building to final users. The lease will usually contain provision for rental and service charges: some of the running costs will be passed on to the user, but the remainder will be met by the level of expenditure covered by the service charge. Improvements in running costs even at the expense of higher capital costs, will make the building more attractive to users. In a reasonably competitive market it follows that rental income will increase as a consequence of reduced running costs. Further, if the additional expenditures are justified by use of whole life techniques, the increased rental will be sufficient to cover the cost of efficiency improvements. A trend in the USA is towards the user paying a rental price that includes the rent and all running costs of the facility; no separate service charge is levied on the tenant. Minimising the running costs on the facility then becomes a very positive objective.

Similar arguments apply for the financial institution. The price it is willing to pay to acquire a facility, by commissioning it or by purchasing from a developer, will be determined by likely rental income and possible capital appreciation. These in turn will be affected by the running costs to be met by the final user.

3.5   The implementation of a whole life appraisal system

The major technical characteristics of the analytical tools on which whole life is based have long been recognised. Whole life appraisal represents a particular application of a classical financial technique, long taught and successfully employed in investment analysis, through which the time-phased costs and revenues attributable to a project over a specified planning period can be evaluated and compared with other methods for achieving a stated objective. As such, the whole life concept is neither new nor complicated, and consists of the following major elements:

- ❏ Identification of an overall time period or whole life, called the period of analysis, applicable to all possibilities being evaluated (within this overall whole life differing whole lives may exist for the various components of the alternative facility or system).
- ❏ Inclusion of all costs and revenues attributable to the project, by time period – including initial investment, recurring costs and revenues and proceeds from ultimate sale or other disposal.
- ❏ Consideration of only those costs and revenues directly attributable to the project decision under consideration.
- ❏ Consideration of the performance characteristics of the asset, such as the maintenance profile, the energy use, and the expected service life.
- ❏ The effects of time, including allowance for
  - ○ the impact of inflation on costs incurred or revenues generated in future years;
  - ○ the fact that money spent or received in the future is worth less than money spent or received today because of reduced interest expense or lost interest income from this money.

The nine steps

The above can be broken into nine steps necessary in implementing a whole life cost approach.

Step 1: Establish the objectives

The single most important step in the analysis is the definition of what the proposed project is intended to achieve, for example:

❑ The budget limit.
❑ The performance requirements.
❑ The design objective.
❑ The risk framework – how much risk the owner/client is prepared to carry.

Typical objectives might be:

❑ To provide $1000\,m^2$ of general administrative space.
❑ To choose between a number of alternative glazing options.
❑ To consider between demolishing and constructing a new school or refurbishing and adapting an existing school.
❑ To consider between alternative sub-bases for a road.
❑ To consider between three options for boilers for a central heating system.

It is essential that the wording of the objective is unbiased, in that it imposes no prior judgements on the best method for achieving the objective. Thus 'build a 4 storey general administrative block' is not an unbiased statement of the first objective above.

Step 2: Establish the constraints

The cost constraints are those beyond which the client is not prepared to go. It might be that the client is only considering owning the facility for five years. Or it might be that the project is in an area of outstanding natural beauty where planning restrictions will only permit limited options.

Step 3: Choice of method

The next step is to determine the range of feasible methods for achieving that objective. Since the ultimate purpose of the whole life process is to assist the decision maker in making resource allocation decisions, it is essential that all realistic possibilities be considered. Occasionally, pre-conceived ideas or administrative constraints (such as an upper limit on initial funding) will tend to exclude certain choices. Nevertheless, all practical possibilities must be analysed. Consider, for example, the situation of two alternatives:

❏   Option 1: Owner-occupation – Building A
❏   Option 2: Commercial lease – Building B

Option 1 is favoured as being the lower whole life cost option. However, a third option – owner-occupation, Building C – has not been considered because its construction cost estimate is felt to be outside capital budget constraints. Further investigation indicates that Building C offers substantial savings in running costs, so much so, in fact, that it is really the lowest whole life cost option. The decision taker must be advised of this option but may still opt for Option 1. This should be done in the knowledge that it is not the most cost-effective solution.

Step 4: Formulate assumptions

Whole life appraisal deals with future expenditure and thus involves elements of uncertainty. A complete factual picture may be impossible to construct and certain assumptions will be necessary in order to proceed with the analysis. For example, it may be necessary to forecast escalation of energy, labour and materials costs. These assumptions must be clearly identified, and where possible accompanied by a statement of the basis for them. Estimates should never be used if factual data are available.

Do not be too prescriptive; component life expectancies are very subjective and depend upon location, use, maintenance and abuse.

Step 5: Identify the period of analysis, the input data for the proposed project, the costs and the whole life

For each possible choice, the performance characteristics must determine the period of analysis, and the design/service life for the whole life of the project and of individual components of that project, and all costs that occur during the entire project whole life. This will often be far from simple to achieve. Seek data from similar past projects, or from manufacturers, or published sources, to provide benchmarks from which to calculate future estimates. Balance performance with cost – remember the purpose is to compare options; so long as a consistent basis is used, the comparison will be valid.

The input data will be the qualitative data about the project being considered. Calculating the input data involves some assumptions, factual information and knowledge.

Step 6: Determine the discount rate to be used for the appraisal, together with the impact of inflation

The real or nominal rate must reflect the client's requirements. Too high a discount rate will swamp decisions about the future.

Step 7: Compare costs and rank the alternatives

This step is the most important element of a whole life approach. Various techniques are available for ranking alternatives, for example, net present value, savings–investment ratios, internal rate of return, or annual equivalent value.

Step 8: Sensitivity analysis

When the results of Step 6 are not demonstrably in favour of one choice, it is advisable to test the sensitivity of the analysis to certain dominant cost factors and assumptions in order to give a complete picture to the decision maker. Sensitivity analysis techniques are discussed in more detail in a later chapter.

The impact of taxation may be an issue that must be taken into account in the sensitivity test.

Step 9: Investigate capital cost constraints

Procedures for whole life appraisal should include a step in which the initial costs of all the recommendations are aggregated to ensure that they do not exceed the funding available. If this constraint is exceeded, trade-off evaluations should be made until the optimum combination of lowest whole life cost within the available funding has been reached. As part of this process, the client should be sufficiently flexible to adjust capital budgets where significant whole life savings are indicated (recall Step 3).

Table 3.1 Whole life cost plan – summary calculations.

Costs	Inflation rate	Inflation-adjusted discount factor	Option 1: Carpet covering Time horizon: 25 years		Option 2: Wood-block covering Time horizon: 25 years	
			Estimated costs	Present value	Estimated costs	Present value
1 Capital costs						
Floor screed (€15/m²)				15 000		15 000
Carpet (€30/m²)				30 000		
Wood-block flooring (€70/m²)						70 000
Total capital costs				45 000		85 000
2 Running costs Operations costs				n/a		n/a
Total operations costs			–	–	–	–
Maintenance costs annual						
Carpet cleaning (€6/m²/pa)	4%	13.085	6 000	78 510		
Wood-block cleaning (€4/m²/pa)	4%	13.085			4 000	52 340
Wood-block refinishing (€10/m²/pa)	4%	13.085			10 000	130 850

Project: Floor covering
Location: New Town
Date: 2004
Discount rate: 5%

	Year	PV factor				
Total maintenance costs annual			5 000	78 510	14 000	183 190
Maintenance/replacement/alterations (intermittent)						
Carpet	10	0.5711	30 000			
Carpet	20	0.3268	30 000			
Total maintenance/replacement/alterations costs				43 865	–	–
Sundries				n/a		
Total sundries				n/a		
Total running costs				122 375		183 190
3 Tax allowances						
Total additional tax allowances						
4 Salvage and residuals						
Total salvage and residuals				none		none
Total net present value of whole life costs				167 375		268 190
Annual equivalent value of whole life costs						

3.6　The output of whole life appraisal

The discussion in this chapter has been somewhat abstract. Thus, while practical applications of whole life will be presented in subsequent chapters, there is some benefit to be gained from a brief illustration of the form that a final whole life cost plan can be expected to take. Such an example is detailed in Table 3.1.

The appraisal

An essential element of whole life appraisal is the factors impacting over the whole life. The factors are:

❑ The period of analysis.
❑ The life expectancy of the components, materials and elements.
❑ How the components will behave when interacting with other components.

The time horizon, sometimes called the period of analysis, over which the study is being conducted, will vary. Typically, this will be the period for which an organisation is expected to operate a particular asset. As a practical and usable definition, however, this is not particularly helpful, since the problem remains of forecasting the probable operational life of a particular facility. Another facet is the design/service life of the facility. A supermarket may only have a 15-year design life, whereas a church would be nearer 100 years. Social housing may require 60 years, and a manufacturing facility 20 years. The design/ service life of roads is more difficult to assess because of changing patterns of use; it is more likely to be between 5 and 10 years before surface replacement is required.

The life of an asset is a complex expression involving many considerations. Theoretically, the lives of various elements should be predicted from observed data on failure, but this type of information is not always available. In addition, life can be extended by periodic maintenance and replacement or may be foreshortened by changing economic, social or legal conditions. These considerations are summarised in Figure 3.6, which shows the sequence of a building whole life for an owner-occupier.

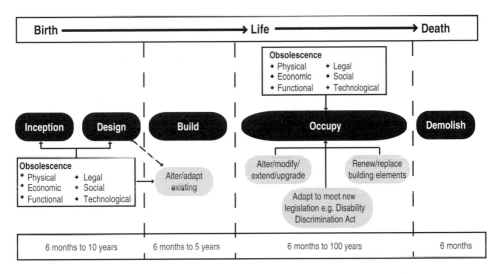

Figure 3.6 Building life cycle.

The life of a facility is difficult to forecast because of:

- Physical obsolescence
- Economic obsolescence
- Functional obsolescence
- Technological obsolescence
- Environmental obsolescence
- Social obsolescence
- Legal obsolescence
- Location obsolescence

What is obsolescence?

Obsolescence occurs as a result of an inability to satisfy changing requirements. Reliable data on obsolescence are rarely available – estimates are based upon experience and intuition.

The physical life

The physical life of the building is the period from construction to the time when physical collapse is possible; in reality most facilities never reach this point, as they are demolished because of economic obsolescence.

Assume a facility with $1000\,m^2$ of floor finishings. The period of analysis is 25 years. The carpet will be replaced every 10 years and the wood-block floor re-finished after 20 years.

The economic life

The economic life is the period from construction to economic obsolescence, that is, the period of time over which occupation of a particular building is considered to be the least-cost alternative for meeting a particular objective. This might occur when the land value of the building is worth more for potential development than the rental income derived from letting the building, or when, for a lessee-occupier, another building becomes more economically desirable.

The functional life

The functional life is the period from construction to the time when the facility ceases to function for the same purpose as that for which it was built. Many clients of the construction sector, particularly in manufacturing industries, require a building for a process that often has a short lifespan. Functional and economic obsolescence are therefore closely interlinked. Functional obsolescence does not normally lead to demolition, as the facility will often be refurbished to suit another function.

Technological obsolescence

Technological obsolescence occurs when the facility or component is no longer technologically superior to alternatives and replacement is undertaken because of lower operating costs or greater efficiency.

Environmental obsolescence

Environmental obsolescence occurs when the facility poses an environmental risk to people or when it no longer meets current environmental legislation.

Social and legal obsolescence

Social and legal obsolescence occur when human desires dictate replacement for non-economic reasons. For example, where some aspects of safety are concerned, replacement becomes essential. Alternatively, where reliability is of the utmost concern, as in hospital buildings, replacement is undertaken.

Location obsolescence

Location obsolescence occurs when an area becomes obsolete because of planning changes. For instance, a new out-of-town shopping centre may cause town centre retail areas to become less viable or even obsolete.

No hard and fast guidelines can be given as to the 'correct' choice of whole life. There is, however, a strong case in favour of choosing economic or functional life. Several points must then be emphasised.

❏ The appropriate economic life will vary with the type of client. An owner-occupier, for example, might be expected to operate a facility in a particular use for rather longer than a rent payer-lessee. When considering the investment for the public sector client a relatively long time horizon, usually commensurate with the physical life, should be used. However, some government departments are given guidelines on specific time horizons to use for investment analysis. Some clients, such as property companies, are looking for a short-term financial gain. In these circumstances the time horizon will equal the clients' economic lifespan for the building, which is the holding period that is expected to maximise speculative profits.

❏ It is probably better to err on the conservative side when forecasting economic life. In addition, the discounting process, by which future costs are converted to their current equivalent, is such that the cost implications of a choice between a 25-year or 30-year whole life will not, in general, be particularly severe.

❏ The period of time over which money has been borrowed to finance the building is not a suitable basis for forecasting the building life. Furthermore, any life calculated purely for taxation and depreciation purposes should not be used for whole life cost calculations.

> Buildings that have been constructed without consideration of maintenance may use 25-year life materials inaccessibly concealed behind 50-year life materials.
> Watson, C. (1999) 'A building is for life not just for sale'.

A distinction must be drawn between the expected service life of a facility and the expected lives of the various components. Whole life techniques should, indeed must, be applied in the choice of both complete facilities and individual components. When looking at a component, however, the whole life is not, in general, the period between acquisition of the component and its eventual replacement. Rather, the component must be seen as an input to the facility within which it is contained. The whole life for the facility component is then the whole life of the facility. The whole life for the carpet covering previously used as an example is not the period within which the carpet will need to be replaced – 10 years – but rather the period of operation of the building for which the floor covering is needed. The carpet will be replaced several times during the life of the facility and an allowance must be made for this.

3.7   Some unresolved problems

In performing the nine steps previously mentioned, a number of further issues will need to be resolved besides those underlying the choice of whole life. First, it will be necessary to identify and collect the data required for effective analysis. This may, however, be far from straightforward. Second, since the analysis takes explicit account of costs over time, there will be estimation problems, made more severe by the lack of an effective historical database.

Inflation poses a similar problem. If all costs can be assumed to escalate at the same rate no great difficulty will arise. If, however, different cost elements typically escalate at different rates, a more complex set of techniques will be necessary. Subsequent chapters will address themselves specifically to these issues.

4 Whole life appraisal at the planning and design stage

4.1 Introduction

The engineering and construction sector has expertise in forecasting construction prices and in manipulating and analysing historical cost and performance data and thus has the ability to apply whole life appraisal techniques. However, the link between the capital and running costs of facilities is missing, as illustrated by Figure 4.1. Furthermore, there is a link between cost information and performance information. Bridging this gap will allow the total cost implications of a design decision to be effectively evaluated at the design stages of a project.

> Research carried out by insurance companies suggests that 40% of latent defects in buildings are design-related, and 40% workmanship related, the remainder arising from component failures. Most importantly, 80% of maintenance costs are built in during the first 20% of the design process.

Figure 4.2 shows the various interests of the parties involved in designing and occupying a facility. At the early design stages the design team is concerned with ensuring that the proposed design meets the client's brief, in terms of function, quality, performance, cost and value for money. As the design develops, the architect, consulting civil, structural and services engineer, and the quantity surveyor must work closely to ensure that the proposed facility is economically sound by achieving a proper balance between capital costs and running costs.

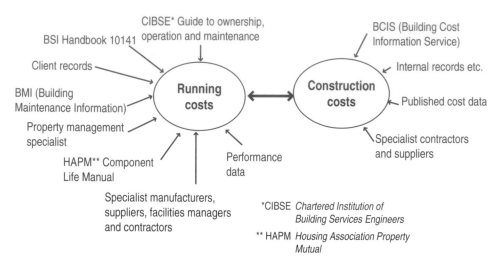

Figure 4.1 The relationship between capital and running costs.

Good intentions

Good intentions to apply whole life appraisal from the inception stage and through the design process are often thwarted by the reality of:

❑ a lack of reliable data on performance;
❑ a lack of time to undertake whole life appraisal;
❑ the need to have the expertise and the resources to evaluate a design from a whole life perspective;
❑ the way the design evolves over time – as more information becomes available so the design is refined and developed.

The design team will seek expert advice about the expected running costs of the project. The personnel most concerned with the running cost aspects are not likely to be involved with the project until the facility is completed and occupied. Furthermore, they will be drawn from a variety of disciplines. This is also complicated by the long time lag between the design stage and the availability of useful data on running costs.

Parties who are likely to have an active involvement in the facility when it is operational can be classified as:

❑ operations staff who are involved in cleaning, security, and the general day-to-day running of the facility;

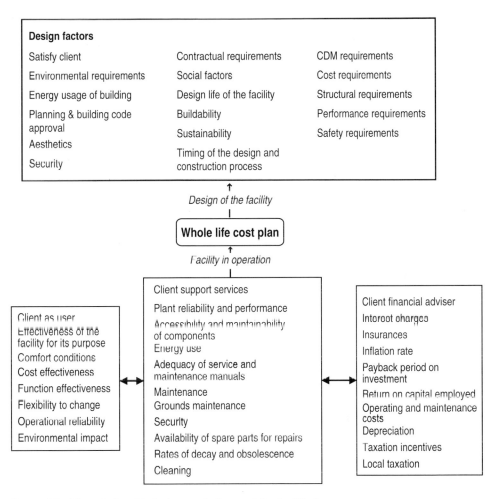

Figure 4.2   The relationship between design and the facility in use.

- ❏ maintenance personnel who are responsible for the planned and corrective maintenance of the fabric and services;
- ❏ management staff who manage the facility (perhaps a facilities' manager),
- ❏ the client's financial staff who provide the technical knowledge on the cost-related aspects of the facility.

Overlaying all these groups is the user of the facility. Clients are interested in using the facility for its design purpose. They often regard the commissioning of a new or newly adapted existing facility as a capital cost process and the running costs of the facility as a revenue cost process. Both public and private sector clients have historically treated these as separate cost centres, partly because of the loan arrangements in the public sector, and partly because of the different treatment of capital and revenue expenses for taxation purposes in the private sector.

A further human factor that adds to the complexity is that a number of assumptions must be made about certain items for the estimated running costs, for example, the period of occupancy and use, the maintenance cycles of certain elements, and the long-term rates of inflation. The design team has no control over many of the issues. It is essential that the assumptions underlying whole life appraisal should be clearly stated and qualified to avoid false security in the precision of the numbers that may be used in an appraisal.

4.2   The components of whole life appraisal

There are several distinct applications of a whole life approach in the construction sector, derived from six questions.

Question 1   What is the total cost commitment of the decision to acquire a particular facility or component over the time horizon being considered?

Question 2   What are the short-term running costs associated with the acquisition of a particular facility or component?

Question 3   Which of several options offers the lowest total whole life cost?

Question 4   What are the running costs and performance characteristics of an existing facility/asset?

Question 5   How can the running costs on an existing facility be reduced?

Question 6   For a build-operate-transfer/concession project – how can the future costs be estimated at the design phase, and what is their reliability?

Each of these questions can be associated with one of four components of an overall whole life approach:

(1) Whole life cost planning and appraisal: applied to Questions 1, 3 and 6.
(2) Full year effect costs: applied to Question 2.

(3) Whole life analysis: applied to Question 4.
(4) Whole life cost management: applied to Question 5.

At their simplest, whole life cost planning and appraisal and full year effect costs are used at the design phase. Whole life cost analysis and cost management are used in the occupancy phase when the facility is in use.

Whole life cost planning and appraisal (WLCPA)

The first use of WLCPA is to identify the total costs of the acquisition of a facility or an individual element. It takes explicit account of initial capital costs and subsequent running costs, and expresses these various costs in a consistent, comparable manner by applying discounting techniques over a time horizon. For example, should a bridge be repaired and refurbished or should it be demolished and replaced by a new bridge?

The second use of WLCPA is to facilitate the effective choice between various methods of achieving a given objective. Indeed, it can be argued that this is the most important aspect of a whole life approach since it will rarely be the case only one possibility is considered. Rather, a choice will have to be made between a number of competing options: which design should be chosen? Should there be tiles or carpet? Should the roof be pitched or flat? What road surfacing should be used?

These options are likely to exhibit different initial capital cost and subsequent running cost profiles. WLCPA provides a set of techniques to convert this diversity of costs to a single consistent measure of cost effectiveness that makes it easy to compare the various options. The output of WLCPA and its interpretation in this second use can be illustrated by considering in more detail the floor-covering example discussed previously.

WLCPA deals with the planning of future costs. The same principles apply to WLCPA as to capital cost planning, where the design team need to have cost targets. An estimated cost target will be set for each of the chosen categories in the WLCPA, which provides a constraint and a measure against which design solutions can be compared. Unlike capital cost planning there is no lowest tender against which to measure the estimating performance. There will be a long time lag before useful running cost data are available for analysis. The temptation must be resisted to manipulate the WLCPA figures to produce a desired result of low running costs.

WLCPA involves collecting and manipulating data from a variety of sources. It is important that a consistent approach be used in the measurement and presentation of data. WLCPA does not require any more measurement information than is currently required in capital cost planning. Unless clients can see considerable financial benefits in WLCPA, they will never use the techniques. This means that the professional service should not involve collection of extensive data that are not readily available to the design team.

WLCPA must be objective, comprehensive in scope, responsive to alternative demands, and accomplished in a timely manner. WLCPA has, of course, the ability to handle both initial and continuing costs, reducing them to a common denominator that can be used as part of the decision process. It follows, naturally, that if a decision has no continuing cost consequences, it makes no sense to undertake a whole life cost plan and appraisal.

Full year effect costs (FYEC)

Sometimes clients will wish to know, at the design stage, the actual running costs of a proposed facility in the short term. Normally an estimate of the running cost for a period of one to three years is provided. This estimate is the FYEC. Future costs are not discounted, but allowance is made in the calculations for the effect of inflation. In essence the client is given information on the likely short-term running costs of a design choice. Many public and private sector clients will use FYEC for budgeting purposes. In practical terms, FYEC is the estimated real expenditure for either new or existing facilities, usually expressed as an annual amount.

Whole life cost analysis (WLCA)

WLCA involves the collection of information on the running costs and performance of facilities in use. In order for the cost data to be meaningful, WLCA must be linked with details about the physical performance, and qualitative characteristics. Figure 4.3 shows the types of information collected within the four data areas.

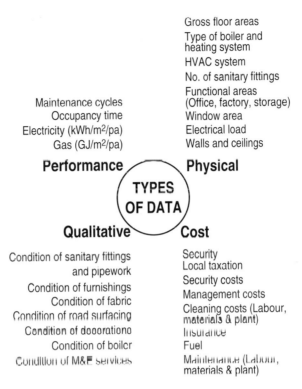

Gross floor areas
Type of boiler and heating system
HVAC system
No. of sanitary fittings
Functional areas (Office, factory, storage)
Window area
Electrical load
Walls and ceilings

Maintenance cycles
Occupancy time
Electricity (kWh/m²/pa)
Gas (GJ/m²/pa)

Performance   **Physical**

TYPES OF DATA

Qualitative   **Cost**

Condition of sanitary fittings and pipework
Condition of furnishings
Condition of fabric
Condition of road surfacing
Condition of decorations
Condition of boiler
Condition of M&E services

Security
Local taxation
Security costs
Management costs
Cleaning costs (Labour, materials & plant)
Insurance
Fuel
Maintenance (Labour, materials & plant)

Figure 4.3 Types of data.

The main use of WLCA is as a management tool intended to identify the actual costs incurred in operating buildings (or, indeed, any durable assets). Since WLCA forms part of overall cost management, it should not be thought of as an end in itself. Rather, by identifying the main items of expenditure incurred in the occupation and maintenance, WLCA will generate an historical database. This database can then be used to highlight areas in which cost savings might be achieved in the design of new buildings, in the operation of existing buildings, and in the choice of individual building components.

The primary objective of WLCA is to relate running cost and performance data and to provide feedback to the design team about the running costs of occupied buildings. In order for this to be effective a system must be developed to enable the data to be collected in a structured fashion. In the case of a new facility, many of the physical and performance data will be available to the design team. However, where an existing facility has been occupied for some considerable time it is likely that the data will not be collected in a fashion suitable for WLCA. The client is unlikely to keep records on the performance and qualitative aspects of his facility. Information is likely to be available in the form of invoices on cost areas such as fuel, local taxation and insurances. It is probable that the occupied facility will need to be measured and calculations undertaken to ascertain the physical performance.

In simple terms, for a building the following questions at least should be asked:

❑   What is the facility type?

And then, for each facility:

❑   Where is it located?
❑   Are there any drawings available?
❑   What is the breakdown of the functional floor area?
❑   What is the general construction?
❑   When was it built?
❑   What running cost information is available?
❑   What is its general condition?
❑   What is the condition of the individual elements (fabric, frame, finishes, etc.)?
❑   What are the periods of occupancy?
❑   What is the maintenance policy?
❑   What performance information is available?
❑   Has the facility been modernised?
❑   What type is the heating installation?

Quite clearly, a much more extensive list of questions could be detailed, but a major consideration is the desire for simplicity. It is pointless to capture data that will not be used in some meaningful fashion. Simplicity is paramount in the design of any system.

The durability of any component depends upon its inherent quality, the environment in which the component is installed, the pattern and intensity of use, local deterioration agents, and the maintenance regime.

Computers help in the storage and retrieval of information. IT systems across an organisation help to capture data on maintenance, energy, utilities and so on. Furthermore, many as-built drawings are stored in an electronic format.

WLCA deals with historical costs and does not involve discounting. When a number of WLCAs have been undertaken for different projects it must be remembered that they relate to cost data for different facilities, in different locations, with different occupancy, at a fixed period of time. No two facilities will have identical running costs, nor will the running costs for any facility be the same from year to year.

Many possible benefits will be lost if the facility's performance and ownership are not monitored throughout its life. It is important to the design team that there is feedback on the cost and performance of the building. In practical terms, this has proved difficult in the past as the design teams must be paid a professional fee to undertake their work.

Whole life cost management (WLCM)

WLCM is a derivative of WLCA. It identifies those areas in which running costs detailed by WLCA might be reduced, either by a change in operating practice, or by changing the relevant system.
WLCM is intended:

❑ To establish where performance differs from the WLCPA projections, why the differences occur, whether they are significant, and whether current performance could be altered.
❑ To make recommendations on more efficient utilisation
❑ To provide information on asset lives and reliability factors for accounting purposes.
❑ To assist in the establishment of a maintenance policy for the facility.
❑ To give taxation advice on building-related items.

Essentially, WLCM is designed to answer the following types of question:

❑ Should the floor covering be changed from tiles to carpet?
❑ Should the office area windows be double glazed to reduce heating costs?
❑ What should be the maintenance/replacement period for the heating/hot water?

In this sense, WLCM is to existing facilities and systems what WLCPA is to new facilities. It is worth restating, that this is one of the most important areas for the application of whole life techniques. Facilities may be expected to endure and be used for many years, during which time design decisions may need to be changed and new options considered.

WLCM will also consider the residual life and the potential for continued service, and whether it is time to decommission because the costs have become too high. Account will be taken of the dismantling and disposal costs and any opportunities for recycling.

4.3   Whole life costing and cost management

In order for whole life appraisal to be adopted, it must prove itself to be economically viable for clients, and professional advisers. Furthermore, it must be seen as a practical tool that enhances existing professional skills and relates in some way to the tasks currently performed.

The temptation must be resisted to dismiss whole life appraisal techniques as being merely costs-in-use techniques with different terminology. Whole life costing brings together the costs-in-use concept and links it to the formal structure of cost planning and long-term ownership costs.

A whole life costing system

To convert theoretical concepts into practical reality requires a formal system. In the same way that cost planning uses elemental categories, so the total cost approach to a project uses whole life categories. Any system of whole life categories is merely a filing system; a series of pigeonholes designed as reminders of possible costs to be included. Whether these are initial capital costs or costs incurred on a continuing or cyclical basis during the life of a facility, they are all costs that arise from, and are affected by, design decisions.

The major whole life cost categories are as follows:

❑   Capital costs (including land and construction costs).
❑   Operations costs (including cleaning, energy, etc.).
❑   Maintenance costs (annual).
❑   Maintenance, replacement and alterations costs (intermittent).

❏   Sundries.
❏   Salvage and residuals.

The concept of levels for whole life cost

The techniques involve the manipulation of large amounts of data. These data can be grouped into a hierarchical structure such as the one shown in Figure 4.4.

The data structure is analogous to the hierarchical library classification based upon the traditional 'tree of knowledge'. Items are classified to successive subdivisions by consistent reference to a single common attribute at each stage. When this concept is related to whole life cost categories it is easiest to think in terms of levels. For example, Figure 4.5 illustrates a structure of levels for maintenance. At Level 1, maintenance as an all-embracing item for the whole building is considered. Level 2 looks at one element associated with maintenance, in this case the finishes, while Level 3 further subdivides this element into the appropriate types of finishes. The whole concept should be self-explanatory; what is important is that the levels can fit a number of tasks associated with WLCA, WLCPA and WLCM.

If a very early design were being evaluated for a WLCPA it is unlikely that any analysis beyond Level 1 would be undertaken. The maintenance item would be analysed at Level 2 as more information becomes available. The point of entry into the levels is dependent solely upon the extent of the information available. Levels 5 and 6 are mainly involved in structuring a maintenance

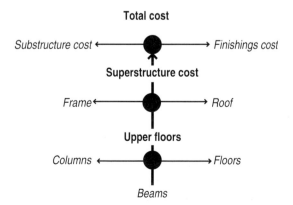

Figure 4.4 Hierarchical data structure.

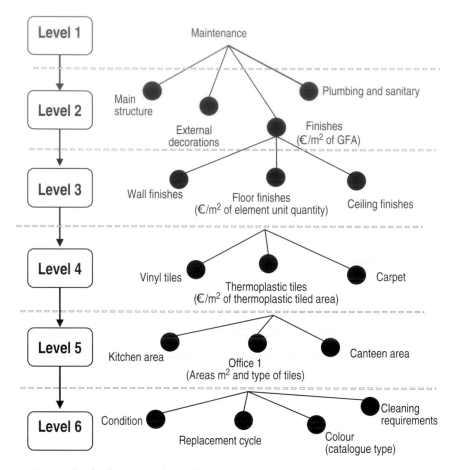

Figure 4.5 Levels of information for maintenance costs.

management programme. For example, at Level 6, by using an effective WLCM, it is possible to identify when the floor tiling in an office is due for replacement.

The problem must be seen in the context of what it would cost to capture this information and why the information is needed. All clients are involved in expenditure on the running costs of their facilities, and therefore the expenditure must be budgeted and planned in some way. Hence the need for a structured approach to whole life appraisal.

In developing the items in the various levels, attention has been paid to the Building Maintenance Cost Information Service and Building Cost Information Service categories, and to the Chartered Institute of Public Finance and Accounting and Department of the Environment maintenance categories.

WLCPA may be used at any stage in the design process. This aspect of whole life costing is all about decision making, whether this is for the selection of alternative glazing types or for establishing the running costs at the inception stage of a complete building. The technique can be useful even if it is not used throughout the design process. The relationship between the levels and WLCA, WLCPA and WLCM is shown in Figure 4. 6. The terms early, brief, budget and detailed have been used to expand the notion of the purpose of WLCA, WLCPA and WLCM.

Whole life cost categories

The table in the appendix at the end of this chapter details the whole life categories developed from Level 1 to Level 3. This appendix has a number of uses:

- It can act as a checklist of items.
- It can form the basis of categories for a WLCPA or WLCM when used in conjunction with measurement information.

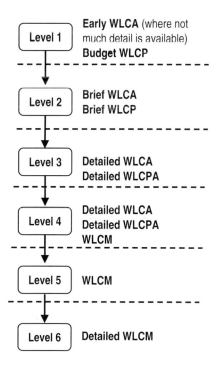

Figure 4.6 The relationship between the levels.

❏ Cost information can be recorded under the various categories when compiling a WLCA.

The list is more suggestive than exhaustive. There are numerous items that have not been included, mainly because it is not feasible to cover every eventuality in one comprehensive list.

Many costs will be incurred annually, but some will be intermittent. For example, under the maintenance section there will be planned/preventive maintenance, corrective maintenance (repairs), upgrading work, adaptation and alteration work. The list must, therefore, be modified to suit each user. Some owners may wish to go down to the level of detail that identifies repairs caused by vandalism, whereas other owners may require only maintenance costs at Level 1.

4.4   Appendix to Chapter 4

Whole life categories developed from Level 1 to Level 3.

Level 1	Level 2	Level 3
1. Capital costs	1A Land	
	1B Fees on acquisition	B1 Legal B2 Estate agents B3 Solicitor B4 Stamp duty B5 Rights of way B6 Rights of light B7 Party wall awards B8 Other
	1C Design team professional fees	C1 Architect C2 Town planner C3 Quantity surveyor C4 Structural engineer C5 Civil engineer C6 Mechanical services engineer C7 Electrical engineer C8 Landscape architect C9 Interior designer C10 Graphic designer C11 Project manager C12 Clerk of works C13 Building surveyor C14 Land surveyor

Level 1	Level 2	Level 3
		C15 Energy consultant C16 Other
	1D Demolition and site clearance	
	1E Construction price for building work	
	1F Cost of statutory consents	F1 Outline and detail planning approval F2 Building regulation approval F3 Listed building consent F4 Conservation area consent F5 Other
	1G Finance for land purchase and during construction	G1 Short term G2 Medium term G3 Long term G4 Other
	1I Capital Gains Tax 1J VAT 1K Furnishings	K1 Carpets K2 Curtains K3 Furniture K4 Other
	1L Removal charges	
	1M Disruption cost (loss of productivity and profits)	
	1N Other capital costs (e.g. plant and equipment)	
	1O Commissioning expenses	
	1P Decanting charges (cost of any temporary accommodation)	
	1Q Other	
2. Finance costs	2A Finance during period of intended occupation	A1 Short term A2 Medium term A3 Long term A4 Other
	2B Loan charges (public sector)	

(Continued)

Level 1	Level 2	Level 3
3. Operation costs	3A Fuel (where possible, apportion fuel bill to appropriate categories, e.g. gas, electricity, oil, etc.)	A1 Heating A2 Cooling A3 Hot water A4 Ventilation A5 Lifts, escalators, conveyors A6 Lighting A7 Building equipment and appliances A8 Special user plant and equipment A9 Other
	3B Water and sewerage charges	B1 Water B2 Sewerage and drainage
	3C Cleaning	C1 Internal surfaces C1a User C1b Circulation C2 External surfaces C2a Windows C2b External fabric C3 Lighting C4 Laundry and towel cabinets C5 External works C6 Refuse disposal C7 Chimneys and flues C8 Other
	3D Local taxation	D1 General D2 Vacancy D3 Other
	3E Insurances	E1 Property E2 M&E and combined engineering E3 Boilers E4 Electric motors and pumps E5 Fixtures and fittings E6 Public liability E7 Employer's liability E8 Loss of profit or rent receivable E9 Special perils E10 Lifts, sprinklers and boilers E11 Other
	3F Security and health	F1 Security services F2 Pest control F3 Dust control F4 Other

Level 1	Level 2	Level 3
	3G Staff	G1 Porterage G2 Caretaker G3 Commissionaire G4 Lift attendant G5 Gardening G6 Uniforms G7 Other
	3H Management and administration of the building	H1 Facilities manager H2 Plan manager/engineer H3 Building management consultancy fees H4 Telephone charges H5 Stationery and postage H6 Other
	3I Land charges	I1 Ground rent I2 Main rent I3 Easements I4 Other
4. Mainten-ance costs	4A Main structure	A1 Substructure A2 Frame A3 Upper floors A4 Roof structure/covering and rainwater drainage A5 Stair structure/finish/balustrade A6 External walls A7 Windows, external doors and ironmongery A8 Internal walls and partitions A9 Internal doors and ironmongery A10 Other
Decorations	4B External decorations	
	4C Internal decorations	C1 Wall C2 Ceiling C3 Fittings C4 Joinery C5 Other
	4D Finishes/fixtures/fittings	D1 Internal walls D2 Internal floors D3 Internal ceilings D4 Internal suspended ceilings D5 Fixtures

(Continued)

Level 1	Level 2	Level 3
		D6 Fittings public liability
		D7 Curtains and finishings
		D8 Other
Plumbing, mechanical and electrical services	4E Plumbing and sanitary services	E1 Sanitary appliances
		E2 Services equipment
		E3 Disposal installation/internal drainage
		E4 Hot and cold water services/ mains water
		E5 Other
	4F Heat source	F1 Boilers, controls, plant and equipment
		F2 Fuel storage and supply
		F3 Other
	4G Space heating and air treatment	G1 Water and/or steam
		G2 Ducted warm air
		G3 Electricity
		G4 Local heating
		G5 Other heating systems
		G6 Heating with ventilation
		G7 Heating with cooling
		G8 Solar collectors
		G9 Heat pumps
		G10 Other
	4H Ventilating systems	H1 Ventilation supply
		H2 Kitchen extract
		H3 Fume extract
		H4 Dust collection
		H5 Smoke extract
		H6 Car parking extract
		H7 Other
	4I Electrical installations	I1 Source mains
		I2 Power supplies and lighting
		I3 Lighting fittings (inc. re-lamping)
		I4 Emergency lighting
		I5 External lighting
		I6 Other
	4J Gas installations	J1 Town and natural gas services
		J2 Distribution pipework to equipment
		J3 Other

Level 1	Level 2	Level 3
	4K Lift and conveyor installations	K1 Lift installations K2 Escalators K3 Hoists K4 Other
	4L Communication installations	L1 Security and fire installations L2 Visual and audio L3 Telephones L4 Other
	4M Special/protective installations	M1 Fire protection M2 Refrigeration equipment M3 Kitchen equipment M4 Laundry equipment M5 Incinerators and flues M6 Water heaters M7 Hand driers M8 Window cleaning equipment M9 Refuse disposal equipment M10 Water pumps M11 Lighting protection M12 M&E equipment associated with occupant M13 Specialist equipment for computer installation M14 Dock levellers to loading bays M15 Sewer pumps M16 Other
External works	4N External works	N1 Repairs and decorations N2 Roads and paved areas N3 Boundaries N4 External services N5 Drainage N6 Fencing N7 Other
	4P Gardening	
5. Occupancy costs		
6. Sundries	6A Energy conservation measures 6B Equipment associated with the building occupier's occupation	B1 Safes B2 Racking to a warehouse B3 Other
	6C Internal planting	

(*Continued*)

Level 1	Level 2	Level 3
7. **Salvage and residuals**	7A Resale value	A1 Building A2 Land A3 Plant and equipment A4 Other
	7B Related costs	B1 Demolition and site clearance B2 Disposal fee and charges B3 Other
	7C Capital Gains Tax	
	7D Disposal costs	

5 Discounting and the time value of money

5.1 Introduction

> The selection of an appropriate discount rate for whole life cost calculations is crucial as it can 'swamp' all other decisions made in the appraisal

The four components (WLCA, WLCPA, WLCM, FYEC) of a whole life approach have one essential common feature, cash flows that arise at different times. A set of techniques is required that will convert future cast outlays to their current equivalent. This chapter outlines such a set of techniques and shows how they can be used.

At this practical level, whole life appraisal is an application of *investment appraisal*. It is, therefore, useful to show why one commonly used method of investment appraisal, payback period, is not appropriate for whole life cost.

> Money today is not the same as money tomorrow. A technique that can express future costs and future revenues into present value is called 'discounting'.

The essence of investment appraisal using the payback period is to identify those options for which revenues most quickly cover the initial capital costs. A first complication in using this technique for whole life cost comparisons is that it will be necessary to generate time series of revenues rather than costs.

Ways can be found for doing this, but nevertheless several other fundamental failings of payback period appraisal remain. These are best identified by example.

Example 5.1

Two design options for a replacement heating system with cost profiles are shown as follows.

Option A

Installation cost	€20 000
Annual savings on existing heating costs	€5000
Expected life of new system	5 years

Option B

Installation cost	€28 000
Annual savings on existing heating costs	€5500
Expected life of new system	12 years

The payback periods are:
Option A = 20 000/5000 = 4 years
Option B = 28 000/5500 = 5.1 years

Option A would be chosen on the payback criterion, but a strong argument can be made in favour of Option B. The essential points are, first, payback ignores all costs and revenues outside the payback period. Thus no account is taken of the fact that Option A will have to be replaced after five years, with consequent additional installation costs. Second, as will be clarified later, a sum of money received or spent today is not equivalent to the same sum of money received or spent next year, or in some years' time. This feature, *the time value of money, is* ignored by payback period methods of investment appraisal.

It must be concluded that payback period is not an adequate method for whole life cost purposes. A method of appraisal is needed which takes account of all present and future cost flows and expresses these costs in a uniform, time-independent manner.

5.2 Discounting

This section begins by investigating in more detail the proposition that 'money today' is in some sense different from 'money tomorrow' and then shows how present and future money flows can be expressed in common terms.

For example, an individual who is given the choice of receiving €1000 now or in a year's time would clearly choose to receive the money now. On the other hand, if offered the choice of settling a €1000 debt now or next year he would choose to pay the €1000 next year. In the former case some return can be expected on the investment, such as the return on a bank or building society deposit, while interest payments can accrue on the money retained. The assumption is that there is no interest payable on the debt.

Given that money at a future date is not equivalent to the same sum of money now, a central problem in whole life costing is to reduce cash expenditures and receipts that arise at different points in time to a common base. It is, therefore, necessary to identify a meaningful 'exchange rate' between money now and money at a future date. This exchange rate is referred to as the *time value of money*.

Consider the case in which an individual, having received €1000 now, can reinvest this sum at a net-of-tax risk-free return of 5% per annum compound by depositing it in a bank. At the end of the first year the €1000 will have grown to €1050, after two years it will have grown to €1102.50 and so on. It can be assumed that this 5% return represents the individual's best available use for the additional funds. In such a case he should value €1050 in a year's time, €1102.50 in two years time, and so on, as equivalent to €1000 now. Put another way, in this example the investor is exchanging money one year hence for money today at a rate of €1.05 for €1. Thus any sum €x to be received one year hence would be valued by him at only €x/1.05 today. Similarly, €x received two years hence would be valued at €x/1.1025 or €x/1.05^2 today. In general, €x received in *t* years' time with rate of interest *r*% is worth today, a present value (PV) of:

$$\text{PV of } x = \frac{\text{€}x}{(1+r)^t} \tag{5.1}$$

The present value factor (PVF) by which a cash flow of €1 in year t can be expressed in its present value is from equation (1):

$$PVF = \frac{1}{(1+r)^t} \tag{5.2}$$

This process by which future money is converted into its equivalent in present money is called *discounting* and derives its name from the fact that future money is being reduced (discounted) to its current money equivalent, i.e. to its present value. The discount rate, the rate at which future money is discounted, is conventionally taken to represent the time value of money. Figure 5.1 shows the effects of alternative discount rates on the present value of €1000 received at varying times in the future.

As can be seen from Figure 5.1, the choice of discount rate will have a significant effect on present value calculations. Consideration must, therefore, be given to the criteria on which the discount rate should be chosen. This is discussed in some detail below. However, in order to make that discussion more meaningful, it is first necessary to illustrate the main principles of discounting.

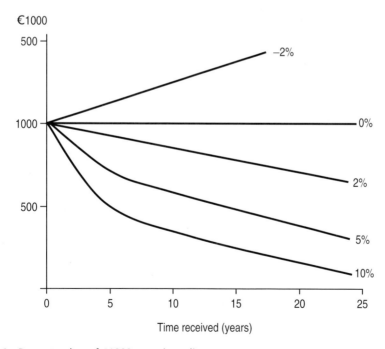

Figure 5.1 Present value of €1000 at various discount rates.

Nominal and real rates

Discount rates may be expressed in:

❑ nominal (market) terms, where both the effects of general price inflation and the real earning power of money invested over time is reflected; or

❑ real terms, where only the real earning power of money is reflected and the effects of inflation are not included.

Discounting methods

The ideas underlying discounting and the main components necessary for effective whole life costing using discounting techniques are best illustrated by a series of simple examples. These examples will illustrate different aspects of discounting methods, but they have one common feature that must be noted at this point. Since the objective of discounting is to produce a present value it follows that this present value will relate to current prices. It is to be expected, however, that costs will escalate over time – it was noted in previous chapters that this is one major reason why a whole life cost approach is now of vital concern to the construction sector. When discounting future cash flows, therefore, account must be taken of the effects of inflation. The principles on which this should be done are discussed in detail below. It is sufficient at this stage to note the simple arithmetic calculations that should be made prior to applying the discounting techniques. Consider a situation in which costs associated with a particular project are expected to escalate at 10% and the long-term market cost of borrowing the finance for that project is 15.5%. Then the net of inflation discount rate is given by:

$$\left(\frac{1.155}{1.10} - 1\right) \times 10 = 5\%$$

More generally, if the expected escalation (inflation) rate is $i\%$ and the long-term market cost of capital is $d\%$, the net of inflation discount rate is given by:

$$r = \left(\frac{1+i}{1+d} - 1\right) \times 100$$

This principle is illustrated by Example 5.2:

Example 5.2

It is projected that an additional €10 000 investment in a building's heating and ventilating system will reduce annual fuel costs (at current prices) by €2000. If it is expected to own the facility for five years, and the additional investment can be expected to increase the resale value by €2000 would the investment be worthwhile?

(It is assumed that the rate of Corporation Tax is 50% and that the investment qualifies for 100% first year capital allowance and that this allowance will be taken totally in the first year.)

The total net of tax cash saving is €7000, but this is received over a period of five years, and so must be converted, i.e. discounted, to its present value. On the assumption that the investment can be financed by borrowing at 15.5% interest compounded annually and that fuel costs are expected to escalate by 10% per annum, the effective discount rate (allowing for inflation) is 5% (1.155 /1.10 = 1.05), and the present value of the net-of-tax fuel costs savings and increased resale value is:

$$\frac{€1000}{1.05} + \frac{€1000}{1.05^2} + \frac{€1000}{1.05^3} + \frac{€1000}{1.05^4} + \frac{€1000}{1.05^5} + \frac{€1000}{1.05^6} =$$

$$952 + 907 + 864 + 784 + 1568 = €5898$$

while the current net-of-tax cash outlay is €5000. The net present value (NPV) of the investment is, therefore, €898. Since this is positive, the investment is worthwhile.

This example can be used to illustrate several other features of discounting. First, the example is a particular illustration of a general principle. The NPV of any stream of costs and revenue occurring over a whole life of N years is given by the general formula:

$$NPV = (R_0 - C_0) + \frac{(R_1 - C_1)}{1 + r} + \frac{R_2 - C_2}{(1 + r)^2} + \ldots + \frac{(R_N - C_N)}{(1 + r)^n}$$

$$= \sum_{t=0}^{N} \frac{R_t - C_t}{(1 + r)^t} \tag{5.3}$$

Second, the calculations can, and preferably should, be presented in tabular form as in Table 5.1.

Table 5.1 Discounted cash flow: example 2.

| | Cash flows | | | | | | | Present value | | |
Year	(1) Outflow	(2) Tax component	(3) Inflow	(4) Tax* component	(5) Net (1 + 2 + 3 − 4)	(6) Present value factor @ 5%	(7) Annuity factor @ 5%	(8) Present value I (5 × 6)	(9) Present value of annuity	(10) Present value II
0	−10000	5000	−	−	−5000	1.000		5000		−
1	−	−	2000	−1000	1000	0.952		952		5000
2	−	−	2000	−1000	1000	0.907		907		
3	−	−	2000	−1000	1000	0.864	4.329	864	4329	4329
4	−	−	2000	−1000	1000	0.823		823		
5	−	−	4000	−1000	3000	0.784		2352		1568
Net present value								−898		+897

The third feature is illustrated in Table 5.1. All cash flows are assumed to arise at the end of the year in which they are incurred or received. Thus the fuel cost savings are assumed to accrue to the firm at the end of years 1, 2, etc. This assumption is not vital. An identical analysis could be performed assuming quarterly, monthly or even daily expenditures and receipts, or on the assumption that all cash flows occur in mid-year. The complications so introduced to the calculations, however, do not in general justify the additional accuracy gained thereby. If the viability of a project is significantly affected by whether cash flows are annual or quarterly, it should be treated with some suspicion!

The fourth feature of discounting is also illustrated in Table 5.1. Calculations can be simplified, and calculating time saved, whenever a project has a number of years of constant net income. The value of the fuel savings in years 1 to 5 can be treated as a 5-year annuity. Consulting annuity tables, sometimes referred to as 'present value of €1 per annum' tables, the present value of €1 per annum for 5 years at 5% is €4329. This is identical, allowing for slight rounding errors, to the sum of the five present value factors in Table 5.1.

The present value of an annuity (PVA) of €1 for N years discounted at r% is:

$$PVA = \frac{(1+r)^N - 1}{(1+r)^N \times r} \tag{5.4}$$

The annuity factor gives the present value of the annuity at the beginning of year 1; that is, at the base date. This is particularly important where regular net cash flows begin other than in year 1, as in Example 5.3.

Example 5.3

As Example 5.2, except that fuel savings in year 1 are only €1800. Calculations are detailed in Table 5.2.

The annuity factor now refers to an annuity for four years, and since the annuity begins in year 2, gives the present value of the annuity as if it arose at the beginning of year 2. This is exactly equivalent to its arising at the end of year 1. Hence, to get the present value of this sum at the base date, it is merely necessary to multiply it by the one-year discount factor (0.952).

Table 5.2 Annuity beginning other than in year 1 (Example 5.3).

Year	Cash flows Total	Tax component	Net cash flow	Present value factor @ 5%	Annuity factor @ 5%	Present value
0	−10000	5000	−5000	1.000		−5000
1	1800	900	900	.952		857
2	2000	−1000	1000		3546 × .952 = 3376	
3	2000	−1000	1000			
4	2000	−1000	1000			
5	2000	−1000	1000			
	2000		2000	.784		1568
Net present value						**+801**

In the original example, the effects on the project's viability of changes in the key elements of the calculations can be examined. Table 5.3 presents the NPV for Example 5.2 under a series of different discount rates and project lives, and identifies those conditions under which the project should and should not be accepted.

Similar calculations can be performed to illustrate the effects of assumptions regarding the probable fuel savings, resale value, or initial cost of the project. This amounts to a sensitivity analysis of the project. Sensitivity analysis will be considered in more detail below; it is a technique intended to identify those elements of a project that have the greatest impact on the project's viability. In looking at Table 5.3, it is necessary to know only that the project life is at least four years for it to be viable, as long as it is reasonably certain that the projected cash flows are accurate, and that the effective (inflation adjusted) discount rate is 5%.

Table 5.3 Net present value with different discount rates and project lives (Example 5.2).

Project life	Discount rate (%) 2	4	5	6	8	10
3	−232	−447	−549	−647	−835	−1011
4	656	340	192	49	−218	−464
5	1525	1096	897	706	355	33
6	2377	1822	1568	1327	883	483
7	3214	2522	2208	1912	1372	894

The main components of discounting

1   The time stream of costs and revenues

It was noted above that the objective of discounting and of discounted cash flow, when applied to whole life cost, is to express future flows of cash in their present value. A central component of the analysis, therefore, is the estimated time stream of expenditures and receipts. It is imperative that all costs and revenues associated with a particular option be identified. In addition, these costs and revenues must take into account such considerations as the impact of taxation and investment incentives. This should on the whole be straightforward for corporate taxation, but where investment incentives, for example, are available as a reduction in the tax liability of corporate profits, calculations will be somewhat more complex.

The costs and revenues that are appropriate will depend upon the individual study. If alternative design options are being compared, for example the plan shape or number of storeys for a particular building, it will be necessary to perform a full whole life cost for each option prior to comparison. On the other hand, if different floor covering options are being compared, an incremental approach might be adopted. One option could be chosen as the base and, as in Examples 5.2 and 5.3 above, others evaluated in terms of their incremental effects on installation costs, running costs and resale value.

This may sound reasonably straightforward, but will often be quite complicated in practice. For example, different glazing options will affect not only installation and energy costs through their differential effects on insulation, but possibly also costs of maintenance, air conditioning, lighting, and cleaning. Effective evaluation of the time stream of costs and revenues requires, therefore, an equally effective view of the building as a complex and interactive system.

2   The discount rate

The second major component of whole life is the choice of discount rate, that is, the choice of the time value of money. Indeed, it can be argued that the discount rate is one of the critical variables in the analysis, in that the decision as to whether to proceed with a particular investment project will be crucially affected by the choice of discount rate. It has already been seen that the process of discounting future cash to give its present value reduces the present value of future receipts, and reduces the present cost of future outlays. What this implies is that the future cost savings (or future cash receipts) generated by a current cash outlay have to be greater the greater is the discount rate. This is shown in Figure 5.2, which illustrates the annual savings required to justify an extra initial expenditure of €1.

For example, a situation may arise in which a choice has to be made between alternative floor coverings for a new office. These coverings are expected to have the same life of 10 years. However, covering B has a higher purchase price and lower annual cleaning costs than covering A. For covering B to be preferred, the annual saving in cleaning costs as compared with covering A will have to be 12 cents for every additional €1 in purchase price at a discount rate of 5%, but would have to increase to 16 cents for every additional €1 of purchase price at a discount rate of 10%. Hence, the higher the discount rate used in the analysis, the lower the impact on the running costs.

It is beyond the scope of this book to enter into an extensive discussion of the principles upon which the choice of discount

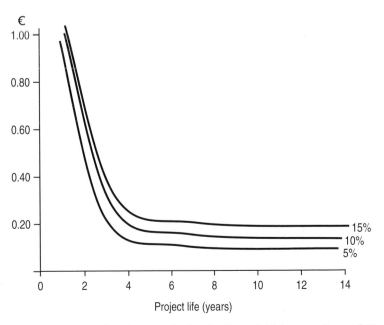

Figure 5.2 Annual savings required to justify an initial expenditure of €1.

rate is based. In the best of all possible worlds the discount rate should be the opportunity cost of capital: the real rate of return available on the best alternative use of the funds to be devoted to the proposed project. A moment's thought will indicate, however, that the 'best alternative use' is not easy, and often impossible to identify. In these circumstances the real (net of inflation) long-term cost of borrowing money in the marketplace – the real rate at which the owner would expect to raise the finance necessary for the project – should be chosen as the discount rate.

It must be emphasised that the appropriate cost of capital is a long-term cost, where the time period refers to the expected life of the project. Whole life cost calculations are applied to long-term investment decisions and cannot be expected to be financed by borrowing in short-term money markets.

Even after it has been decided to use the cost of capital as the discount rate, the appropriate rate will vary with the source of funding. There are three important categories of finance available to companies, which, ranked in descending order of their probable costs are as follows:

❏   Funds from the issue of shares.
❏   Retained earnings.
❏   All flows of fixed interest debt capital.

Since new issues are likely to constitute the most expensive form of funding, and fixed interest capital the least expensive, a company might be expected always to choose fixed interest capital. However, lenders will generally restrict the amount of long-term debt capital that can be raised by the company since this constitutes a prior charge on the company. Similarly, there will be restrictions on the proportion of earnings that a company can retain, if only from the need to maintain a reasonable return to equity holders in the company.

Where the company is forced to adopt a form of funding other than fixed interest capital, problems may arise in identifying the cost of borrowing. If the project is financed from retained earnings, the required rate of return will be less than that required on new issues, since the company can obtain the return required by equity holders from a lower return on retained earnings. This arises because finance provided by retained earnings offers substantial tax savings; specifically, the company need not take into account the tax liability incurred by equity holders on dividend payments. Thus the appropriate rate of discount if funding is solely by retained earnings should be the cost of equity finance adjusted appropriately for tax savings.

Further problems arise where the project is funded by a mix of some, or all, of the three sources of finance identified above. Two main mixes should be considered, as follows.

Retained earnings and fixed interest capital

In this case the appropriate discount rate will be a weighted average of the returns required on these two types of capital. The weights appropriate to calculating this average are the proportions in which the two types of capital will be used to finance the project. Thus, if a company plans to finance a new building by a mixture of 80% retained earnings and 20% debt capital, and if the required returns on these two sources of finance are 11.5% and 4% respectively, the weighted cost of capital would be:

$$(0.8 \times 11.5) + (0.2 \times 4.0) = 10.0\%$$

New issues and fixed interest capital

Assuming that a company has raised the maximum feasible loans on its existing assets and income, further loans can be raised only when an appropriate proportion of new equity finance is provided. The cost of finance in this case will be a weighted average of the returns required on equity and debt capital. Assuming proportions of 80% equity and 20% new debt and required returns of 14.0% and 4.0% respectively, the weighted cost of capital would be:

$$(0.8 \times 14.0) + (0.2 \times 4.0) = 12.0\%$$

It should be noted that if a mix of retained earnings and new equity capital finances the project, the cost of capital should be the cost of the equity capital. The company will, presumably, consider this option only when it is already retaining something approaching the maximum practical proportion of earnings. New issues will not alleviate this position; indeed, a new issue may reduce the proportion of earnings that the company feels able to retain. In this case, therefore, the lower return required on retained earnings is irrelevant, since, if the project under consideration proves to be unacceptable, the reduction in capital requirements will be made in the relatively expensive new share issue source of funds. In other words, the marginal cost of capital in this case is the cost of externally raised equity capital, and this is the standard against which the project should be measured.

The impact of risk up to this point has not been discussed. Investment in government bonds can be assumed to have no risk. Hence, if the bond rate of return is 5%, less an allowance of 2% for inflation, then the real discount rate would be 3%. Investment in equities carries lots of risk with stock markets becoming more volatile. Similarly, investment in property carries risk. Hence, a risk premium discount rate is often applied to whole life appraisal. For example, if the average equity rate of return is 10% and the government bond rate is 5%, the average risk premium discount rate is 5%.

3 Project life

The third major component of discounting is an estimate of the probable service life of the project. Many factors will influence this estimate, and these do not allow the formulation of hard and fast rules. Wherever possible, estimates of probable material and component lives should be obtained from specialist suppliers. In many cases, however, reliance is placed upon historical data and professional judgement.

The effects of variation in project service life are considered below. It is sufficient to state at this stage that the shorter the project life, the more important it is that the estimate of project life be accurate.

5.3   Choosing between alternative options

Ranking projects with identical lives

One of the major purposes of discounting techniques within the context of whole life cost is to allow the decision maker to choose between possible methods of achieving a given objective. Thus a choice may have to be made between different plan shapes, building aspects, lighting options, glazing options, or floor coverings in situations where the options are mutually exclusive: it is not possible to put a five-storey and ten-storey building of equal floor space on the same piece of land. The decision maker must, therefore, be able to use discounting methods to rank the options that are presented.

The method of ranking depends upon how the calculations are presented. Two basic approaches are available if the options being ranked have identical forecast lives: the absolute and incremental approaches. With the absolute approach the NPV for each possibility is evaluated prior to comparison. The incremental approach, on the other hand, takes one option as the base – it does not matter which – and calculates the NPV of the difference between that base and any other. Both methods will give the same ranking provided all options have the same forecast lives.

An example may make this clearer.

Example 5.4

A client can choose between three methods of floor covering for a new office building. All coverings are assumed to have the same lives, forecast as five years, after which they will need to be replaced. They have no resale value. The cost of capital adjusted for inflation is 5%, and installation and running costs at current prices are detailed in Table 5.4. Capital allowances, at 100%, are taken fully in the first year, and Corporation Tax is assumed to be payable at 50%.

Which option should the client choose?

1 Absolute approach

Calculations are presented in Table 5.5.

Table 5.4   Costs (Example 5.4).

		Floor covering option					
		A		B		C	
Costs		Gross	Net of tax	Gross	Net of tax	Gross	Net of tax
Installation		10000	5000	11200	5600	13600	6800
Running	1	1000	500	600	300	400	200
costs in	2	1000	500	700	350	400	200
year	3	1400	700	1100	550	600	300
	4	1800	900	1500	750	800	400
	5	2400	1200	2000	1000	1200	600

Table 5.5   Present value: absolute approach (Example 5.4).

Year	Present value factor @ 5%	Present value of costs* for Option		
		A	B	C
0	1.000	5000.00	5600.00	5000.00
1	0.952	476.00	286.60	190.40
2	0.907	453.50	317.45	181.40
3	0.864	604.80	475.20	259.20
4	0.823	747.70	617.25	329.20
5	0.784	940.80	784.00	470.40
Total present value		8215.80	8079.50	8230.60

* net of tax

2   Incremental approach

Taking option A as the base, net of tax cash flows may be calculated for 'A–B' and 'A–C', and NPV calculated for the hypothetical projects 'A–B' and 'A–C', as in Table 5.6.

Both methods rank the alternatives in the order B, A, C.

It may be noted that with the absolute approach the best, that is lowest cost, option is that with the lowest present value, whereas with the incremental approach, the best option is that with the highest net present value. The apparent difference in ranking procedure arises since in whole life cost, cash flows will typically be costs rather than revenues. In terms of equations (5.3) and (5.4) above, costs will enter the calculations as negative numbers.

It is more convenient in these circumstances to ignore the negative signs and choose that option with the lowest present value, since this is equivalent to the lowest cost.

Table 5.6   Present value: incremental approach (Example 5.4).

Year	Present value factor @ 5%	Cash flows*		Present value for:	
		A–B	A–C	A–B	A–C
0	1.000	−600	−1800	−600.00	−1800.00
1	0.952	200	300	190.40	285.60
2	0.907	150	300	136.05	272.10
3	0.864	150	400	129.60	345.60
4	0.823	150	500	123.45	411.50
5	0.784	200	600	156.80	470.40
Total present value				136.30	−14.80

* net of tax

When using the incremental approach, on the other hand, the hypothetical project 'A–B', for example, incurs costs of installation, but then offers savings (revenues) on running costs. The best option is that with the highest positive NPV: note that the hypothetical project 'A–A' has an NPV of zero. Thus option B offers additional savings with a net present value of €136.30 when compared with option A, while option C incurs additional cost with a net present value of €14.80.

The incremental approach offers no advantages when compared with the absolute approach, since both use exactly the same information. It does, however, present problems when options have different projected lives, since it will be necessary to create hypothetical projects with identical lives before comparison is possible. In addition, it will often be the case that whole life cost techniques will be applied in cases where 'maintain existing system' is one feasible option, for example, when deciding whether to replace a particular building system. It is preferable that costs of the existing system be presented explicitly, rather than that costs for proposed options be presented as changes to existing costs. This will force the decision-maker to detail the assumptions made, and will provide better information with which to make the decision.

For both reasons, it is recommended that the absolute approach be used in whole life cost calculations. One point should then be emphasised. Any resale value for a particular option should be entered as a negative number since it is a deduction from costs. Similarly, if, when a new option such as a new heating system is adopted, the existing system has a resale value, this should also be entered as a negative number in the calculations for the proposed new system.

Ranking projects with different lives

The problem remains of the choice between options with different project lives. How, for example, should an occupier choose between two types of floor covering, one with an estimated life of three years, and the other with an estimated life of five years? One possibility would be to treat the former as five consecutive three-year decisions, and the latter as three consecutive five-year decisions, both totalling fifteen years. If, however, there were a third option with an economic life of seven years, it would be necessary to expand the time horizon to 105 years!

A much simpler procedure is available by calculating the present value of each possibility and expressing this as a uniform annuity or annual equivalent (AE). The least cost option is that with the longest AE.

The annual equivalent is the discounted life cycle cost converted to a uniform annual amount which would pay off the costs over the study period.

In other words, if a particular option has present value of costs equal to PV, and expected life of N years, we wish to find the N year uniform annuity or annual equivalent (AE) which has present value of costs equal to PV. This annual equivalent is identified from the equation:

$$PV = \sum_{t=1}^{N} \frac{AE}{(1+r)^t} = AE\frac{(1+r)^N - 1}{(1+r)^N \times r} \tag{5.5}$$

from which it follows that:

$$AE = PV\frac{(1+r)^N \times r}{(1+r)^N - 1} \tag{5.6}$$

For example, a project may be considered with an economic life of ten years, and PV of costs of €5000 at a discount rate of 5%. Using equation (5.6):

$$AE = €5000 \times \frac{(1.05)^{10} \times 0.05}{(1.05)^{10} - 1} = €647.50$$

Thus, the present value of the cost stream (€5000) is equivalent to a uniform cost stream of €647.50 over the ten-year life of the project at a discount rate of 5%.

Calculation of AE is considerably simplified by using annuity tables. An annuity of €1 per annum for ten years at 5% has present value:

$$PVA = €7.722$$

and comparing equations (5.5) and (5.6) it follows that:

$$AE = \frac{PV}{PVA} = \frac{£5000}{7.722} = £647.50 \tag{5.7}$$

An example may further clarify this principle.

Example 5.5

In the case of the floor covering options of Example 5.4, it may now be considered that option B has a project life of six years and option C a project life of eight years. The inflation-adjusted discount rate is 5%. Tax rates are as in Example 5.4, and net of tax costs for option A are as in Table 5.4 while those for options B and C are as in Table 5.7.

Now option C should be chosen as the lowest cost alternative.

This example should also serve to emphasise why the incremental approach should not be adopted when alternative options have different forecast lives. The incremental approach cannot be applied directly to the data in Table 5.7 since there is no information on the costs of option A, for example, in years 6, 7 and 8. Converting these options to equivalent lives, on the other hand, requires analysis of the projects over a minimum period of 120 years.

5.4   Internal rate of return (IRR)

While the discussion in previous sections has concentrated on discounted cash flow and net present value, some comments are

Table 5.7   Present value calculations (Example 5.5).

Year	Costs* for option A	B	C	Present value factor @ 5%	Present value of costs* A	B	C
0	5000	5600	6800	1.000	5000.00	5600.00	5000.00
1	500	300	200	0.952	476.00	286.60	190.40
2	500	350	200	0.907	453.50	317.45	181.40
3	700	550	300	0.864	604.80	475.20	259.20
4	900	750	400	0.823	747.70	617.25	329.20
5	1200	1000	600	0.784	940.80	784.00	470.40
6		1200	800	0.746			
7			1000	0.711			
8			1200	0.677			
Total present value (PV)					8215.800	8974.700	10350.800
Present value of annuity (PVA)					4.329	5.076	6.463
Annual equivalent (AE=PV/PVA)					1897.850	1768.070	1601.550

* Net of tax

necessary on an alternative method of investment appraisal – the internal rate of return (IRR). The IRR for a project is defined as that discount rate which generates an NPV of zero. Recalling equation (5.3):

$$NPV = \sum_{t=0}^{N} \frac{R_t - C_t}{(1+r)^t}$$

while IRR is that interest rate k which is such that:

$$NPV_k = \sum_{t=0}^{N} \frac{R_t - C_t}{(1+k)^t} = 0 \qquad (5.8)$$

The first limitation of IRR as a method of ranking for whole life cost purposes is that it can be applied only if the incremental approach is adopted: if all cash flows on an option are negative (that is, are costs), no IRR can be calculated.

Many more serious criticisms can be levelled at IRR in comparison with NPV as a method of ranking alternatives. IRR is more difficult to calculate, can give inconsistent recommendations, and contains a logical error in its methodology. To detail these criticisms is beyond the scope of this book. The interested reader can check them in any text on capital investment appraisal (see, for example, Levy, H. and Sarnat, M. (1982) *Capital Investment and Financial Decisions* (2nd edn), Prentice Hall International, especially Chapter 4.). Suffice it to say that there are compelling reasons for preferring present value comparison as the method of appraisal for whole life cost purposes.

5.5   Inflation

Inflation in cost is important. Indeed, it can be argued that one of the main reasons for the growing interest in whole life appraisal has been precisely because certain costs, such as energy and labour, have risen sharply in recent years and called into question the traditional design approach. As a consequence, present value calculations within a whole life framework must be capable of taking inflation into account. Methods for doing so will now be considered in detail.

The simplest method might appear to be to evaluate all cash flows in real terms and discount at an inflation-free discount rate. Thus, if a project is expected to generate costs of €1000 per annum for five years, evaluated at today's prices, if all costs are expected to rise at 5% per annum over the five years, and if the market cost of capital is 10.5%, the simplest approach will be to discount the cost stream at 5%.

It must be accepted that in many instances this will be the best information available at the time of analysis. Indeed, it will often be the case that when a whole life cost plan is prepared there may not even be an adequate estimate of expected escalation rates. Where this is so, adopt an approach identical to that recommended by government with respect to investment in the public sector and apply a test discount rate. Just what should this test discount rate be? Recent analysis indicates that when inflation rates are reasonably low (less than 10%) there is quite a stable relationship between inflation and bank base rate, implying a real discount rate of between 3% and 4%. If no better information is available, therefore, it is recommended that a test discount rate of 4% be used.

The test discount rate should be used only if better information is not available, since it is crucially dependent upon the assumption that inflation will apply equally to all future cash flows on the project. In some cases this may be sufficiently true to be taken as a working approximation, or may be necessary because of lack of information, but in any complex investment project a wide range of different factors – labour, fuel materials – will be involved, the costs of which can be expected to inflate at different rates.

In circumstances where different components can be expected to increase in price at different rates and where adequate information on relative inflation rates can be generated, a much more sophisticated approach should be adopted in incorporating inflation into the analysis. Essentially, what this involves is the evaluation of each component in money terms prior to discounting at a cost of capital inclusive of an allowance for inflation, i.e. the market-generated cost of capital. An example may serve to make this clearer.

Example 5.6

A firm proposing to build a new commercial facility is considering two design options with costs detailed in Table 5.8. These costs are assumed to be net of all tax allowances.

Table 5.8 Hypothetical office costs (Example 5.6).

	Option A	Option B
Construction costs	€1.5 m	€2.0 m
Running costs (current value)		
(a) Energy (heating, lighting, air conditioning)	€100 000	€80 000
(b) Cleaning (labour the only variable)	€40 000	€30 000
(c) Maintenance: Labour	€30 000	€24 000
Materials	€10 000	€8 000
Expected design/service life	15 years	15 years
Resale value (allowing for inflation)	€3.0 m	€3.0 m

Cost of capital		15%
Expected inflation rate Energy		11%
Labour		8%
Materials		8%

NB: All cash flows are in current prices with the exception of the resale value.

Which design option should be chosen? To calculate present values it is first necessary to calculate discount rates for the various costs.

(a) **Energy costs** are expected to inflate at 11% and should be discounted at 15%. This is equivalent to a discount rate of $(1.15/1.11) - 1 = 3.6\%$. Using 15-year annuity factors:
@ 3% = 11.938
@ 4% = 11.118
hence @ 3.6% = $11.938 \times 0.4 + 11.1181 \times 0.6 = 11.446$

(b) **Cleaning costs** are expected to inflate at 8%.
The discount rate is $(1.15/1.08) - 1 = 6.5\%$
The 15-year annuity factor @ 6.5% = 9.410

(c) **Maintenance costs** are expected to inflate at 8%.
The discount rate is $(1.15/1.08) - 1 = 6.5\%$
The 15-year annuity factor @ 6.5% = 9.410

Given these discount rates, present value calculations are as in Table 5.9.

Option A should be chosen since it is the lower cost option.

Example 5.6 shows that so long as the rate of inflation of a particular cost estimate is less than the market discount rate, the principles of present value calculation are precisely those already discussed. All that is needed is calculation of the net of inflation discount rate. If expected inflation of a particular cost

Table 5.9 Present values (€m) (Example 5.6).

	Cash flow		Annuity factor	Present value factor @ 14%	Present value	
	A	B			A	B
Construction	1.5	2.0		–	1.5	2.0
Running costs						
Energy	0.1	0.08	11.446	–	1.145	0.916
Cleaning	0.04	0.03	9.410	–	0.376	0.282
Maintenance						
Labour	0.03	0.024	9.410	–	0.282	0.226
Materials	0.01	0.008	9.410	–	0.094	0.075
Resale value	−3.0	−3.0	–	0.140	0.42	−0.42
Total present value (€m)					**2.977**	**3.079**

element is $i\%$, and the market rate of discount is $r\%$, the net of inflation discount rate the decision-maker should use is:

$$d = \frac{1+r}{1+i} - 1 \tag{5.9}$$

For instance, in Example 5.6 with $d = 15\%$ and $i = 8\%$ the net of inflation discount rate is:

$$\frac{1.15}{1.08} - 1 = 6.5\%$$

Complications arise, however, when the expected inflation rate for particular costs exceeds the cost of capital. The method of calculation presented in Table 5.9 cannot be used for such costs.

It may be assumed that a particular cost, estimated at today's prices as $€C$ per annum, is expected to inflate at an annual rate of $i\%$ over its life of N years. The discount rate is $r\%$. Then the present value of these costs is:

$$PV = \sum_{t=0}^{N} \frac{C \times (1+i)^t}{(1+r)^t} = C \times \frac{\left(\frac{1+i}{1+r}\right)^N - 1 \times \frac{1+i}{1+r}}{\frac{1+i}{1+r} - 1} \tag{5.10}$$

Calculation of the present value in equation (5.10) is considerably simplified by defining a net inflation rate (e) by the formula:

$$1 + e = \frac{(1+i)}{(1+r)}$$

Thus, for example, if expected inflation is 17.3% and market cost of capital is 15%

$$1 + e = \frac{1.173}{1.15} = 1.02$$

then the net inflation rate is 2%. It is then possible to tabulate the present value of €1 per annum at various net inflation rates. These values are detailed in Appendix A to this chapter (see page 92).

The net inflation rate is, in fact, used when the effective (real) discount rate is negative. This can be illustrated by considering a situation in which a client can borrow €1000 to upgrade his building insulation, so reducing heating and ventilating costs. The client has been advised that heating and ventilating costs are expected to escalate at 15% per annum, while he can borrow the €1000 at a market rate of 12% per annum. In these circumstances, if the client upgrades the insulation, the future savings achieved will actually be growing at an effective rate of 2.7%, equivalent to a discount rate of −2.6%. To see why this is so, note that with inflation at 15%, a €1 reduction in fuel costs at today's prices is equivalent to €1.15 in one year's time, and at a discount rate of 12% this has a present value of €1.15/1.12 = €1.027: a growth or net inflation rate of 2.7%. Putting this in terms of discount rates, what is needed is that discount rate r is such that $1/(1 + r) = 1.027$. Solving this equation gives $r = -0.026$, i.e. a discount rate of −2.6%. Appendix A notes the effective (negative) discount rates associated with each net inflation rate. For example, a net inflation rate of 2% is equivalent to a negative discount rate of −1.96%.

As an illustration of these equations, it may be assumed in Example 5.6 that energy costs are forecast to increase at 17.3% per annum. Then the net inflation rate is given by $1 + e = 1.173/1.15 = 1.02$, i.e. e is 2%.

From Example 5.6:

$$C = €100\,000 \text{ for Option A}$$

$$C = €80\,000 \text{ for Option B}$$

Using Equation (5.10) the forecast value of energy costs (in €m) is:

$$0.10 \times \frac{\left(\frac{1.173}{1.15}\right)^{15} - 1 \times \frac{1.173}{1.15}}{\left(\frac{1.173}{1.15}\right)^{15} - 1} = €1.764 \qquad \textbf{Option A}$$

$$0.08 \times \frac{\left[\dfrac{1.173}{1.15}\right]^{15} - 1 \times \dfrac{1.173}{1.15}}{\left[\dfrac{1.173}{1.15}\right] - 1} = €0.411 \qquad \textbf{Option B}$$

Alternatively, consulting Appendix A, with net inflation rate 2% and project life 15 years, the present value of an inflating annuity is 17.639, and the same results are obtained. Note that in these circumstances Option B in Example 5.6 would be the preferred option.

As a general rule, it is to be expected that once inflation is allowed for in the analysis, investment decisions will become much more sensitive to future operating costs. Inflation reduces the effective discount rate. The greater is the expected rate of inflation relative to the money cost of capital, the greater will be the consequent reduction in the effective discount rate.

5.6   Risk, uncertainty and sensitivity analysis

The concluding remarks of the last section hint at one relatively important feature of whole life cost calculations. The majority of such calculations will be based upon assumptions about the costs of fuel, cleaning, maintenance and so on, consequent upon a design decision with an estimated initial capital cost. Forecasts will be made regarding future escalation rates of these various costs, and an estimate will be made of the probable project life.

Clearly, the cash flows, escalation rates and project life cannot be known with certainty. Rather, information will be available only on the likely bounds within which they will lie, or the range of possibilities.

There is now an extensive literature on risk analysis and choice under uncertainty. All that needs to be emphasised here is that estimates should be based upon the best possible information. An important part of any estimate or forecast is the expertise with which they are produced as basic input data are inevitably uncertain.

Simple techniques are available to facilitate identification of the most important, or sensitive, assumptions. In particular, once estimates have been made it is recommended that the results of the whole life calculations be subjected to a sensitivity analysis with respect to those estimates or assumptions felt to be most uncertain.

Sensitivity analysis is a technique that analyses the effects on a project's viability of variations in particular elements of that project. The technique is best illustrated by means of an example.

Example 5.7

A client has to choose between two glazing options, each with uncertain economic life. Costs of these options are detailed in Table 5.10. All figures are adjusted for taxation and there is assumed to be no inflation. The client expects the cost of capital to be between 5% and 8%. Calculation of the present value of these options will result in a complex of results since there is no direct knowledge of either project life or the actual cost of capital. Four basic cases can, however, be identified for each option as in Table 5.11 and the present values for these cases can be illustrated in the graphical form of Figure 5.3.

The graphical analysis of Figure 5.3 allows immediate identification of the variables that have important effects on the choice. It focuses attention on the critical outcomes, and makes the possible range of estimates clear. The final decision resulting

Table 5.10   Cost data (Example 5.7).

Costs (€000)*	Option A	Option B
Installation costs	150	100
Annual costs of energy loss	15–25	25–35
Additional annual cleaning and operating costs	10–15	15–20

* Net of tax

Table 5.11   Possible cash flows for Example 5.7 (€000).

	Option A			Option B		
	Energy costs	Operating costs	Total annual costs	Energy costs	Operating costs	Total annual costs
Case 1	15	10	25	25	15	40
Case 2	15	15	30	25	20	45
Case 3	25	10	35	35	15	50
Case 4	25	15	35	20	20	55

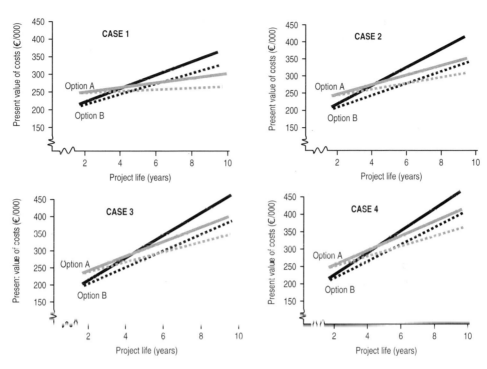

Figure 5.3   Present value cost comparisons (Example 5.7).

from the sensitivity analysis remains a matter of subjective judgement, of course, but the decision taker now has much more solid information on which to base that judgement, and can identify the effects of his decision much more clearly.

No definitive rules can be laid down for decision making in uncertain conditions. Such decisions will inevitably be based on value judgements and, in particular, on the attitude of the individual decision maker. This section has shown, however, that correct presentation of data will be of significant benefit to those who carry the responsibility for the final decision.

5.7   Summary

The aim of this chapter has been to indicate how particular investment appraisal techniques can be applied to whole life appraisal. These techniques are necessary, first, because money today and money tomorrow are not equivalent and second, because whole life appraisal is applicable precisely in those conditions where costs are arising over time.

While there are several methods of investment appraisal, only one is properly applicable to whole life appraisal. Projects should be assessed on the basis of their present values, using a discount rate equal to the real long-term cost of capital to the organisation. Methods such as pay-back period and internal rate of return are not appropriate for whole life cost purposes. Complications will arise if the options being compared have different economic lives, or are subject to inflation, or involve estimates that are uncertain.

Techniques have been presented to incorporate differences in project lives or inflation in the analysis. Problems remain, however, with respect to uncertainty. In these circumstances a sensitivity analysis should be performed. This will identify those assumptions or estimates to which the viability of the project is most sensitive. It will, therefore, improve the information base upon which the eventual choice will be made.

5.8   Appendix A to Chapter 5

Present value rate of €1 at net inflation rate.

Years	1%	2%	3%	4%	5%	6%	7%	8%	9%	10%
1	1.009	1.020	1.030	1.040	1.050	1.060	1.070	1.080	1.090	1.100
2	2.030	2.060	2.091	2.120	2.153	2.184	2.215	2.246	2.278	2.310
3	3.060	3.120	3.184	3.246	3.310	3.375	3.440	3.506	3.573	3.641
4	4.101	4.204	4.309	4.416	4.526	4.637	4.751	4.867	4.985	5.105
5	5.152	5.308	5.468	5.633	5.802	5.975	6.153	6.336	6.523	6.716
6	6.214	6.434	6.662	6.898	7.142	7.394	7.654	7.923	8.200	8.487
7	7.286	7.583	7.892	8.214	8.549	8.897	9.260	9.637	10.020	10.436
8	8.369	8.755	9.159	9.583	10.027	10.492	10.978	11.488	12.021	12.579
9	9.462	9.950	10.464	11.006	11.578	12.181	12.816	13.487	14.193	14.937
10	10.567	11.169	11.808	12.486	13.207	13.972	14.784	15.645	16.560	17.531
11	11.683	12.412	13.192	14.026	14.917	15.870	16.888	17.977	19.141	20.384
12	12.809	13.680	14.618	15.627	16.713	17.882	19.141	20.495	21.953	23.523
13	13.947	14.974	16.086	17.292	18.599	20.015	21.550	23.215	25.019	26.975
14	15.097	16.292	17.599	19.024	20.579	22.276	24.129	26.152	28.361	30.772
15	16.258	17.639	19.157	20.825	22.657	24.673	26.888	29.324	32.003	34.950
16	17.430	19.012	20.762	22.698	24.840	27.213	29.840	32.750	35.974	39.545
17	18.615	20.412	22.414	24.645	27.132	29.906	32.999	36.450	40.301	44.599
18	19.811	21.841	24.117	26.671	29.539	32.760	36.379	40.446	45.018	50.159
19	20.019	23.297	25.870	28.778	32.066	35.786	39.995	44.762	50.160	56.275
20	22.239	24.783	27.676	30.969	34.719	38.993	43.865	49.423	55.765	63.003
21	23.472	26.299	29.537	33.248	37.505	42.392	48.006	54.457	61.873	70.403

(*Continued*)

Years	1%	2%	3%	4%	5%	6%	7%	8%	9%	10%
22	24.716	27.845	31.453	35.618	40.430	45.996	52.436	59.893	68.532	78.543
23	25.973	29.422	33.426	38.083	43.502	49.816	57.177	65.765	75.790	87.497
24	27.243	31.030	35.459	40.646	46.727	53.865	62.249	72.106	83.701	97.347
25	28.526	32.671	37.553	43.312	50.113	58.156	67.676	78.954	92.324	108.182
26	29.821	34.344	39.710	46.084	53.669	62.706	73.485	86.351	101.723	120.100
27	31.129	36.051	41.931	48.986	57.403	67.528	79.698	94.339	111.968	133.210
28	32.450	37.792	44.219	51.966	61.323	72.640	86.347	102.966	123.135	147.631
29	33.785	39.568	46.575	55.085	65.439	78.058	93.461	112.283	135.308	163.494
30	35.133	41.379	49.003	58.328	69.761	83.803	101.073	122.346	148.575	180.943
31	36.494	43.227	51.503	61.701	74.299	89.890	109.218	133.214	163.037	200.138
32	37.869	45.112	54.078	65.210	79.064	96.343	117.933	144.951	178.800	221.252
33	39.258	47.034	56.730	68.858	84.067	103.184	127.259	157.627	195.982	244.477
34	40.660	48.994	59.462	72.652	89.320	110.435	137.237	171.317	214.711	270.024
35	42.077	50.994	62.276	76.598	94.836	118.121	147.913	186.102	235.125	298.127
36	43.508	53.034	65.174	80.702	100.628	126.268	159.337	202.070	257.376	329.039
37	44.953	55.115	68.159	84.970	106.710	134.904	171.561	219.316	281.630	363.043
38	46.412	57.237	71.234	89.409	113.095	144.058	184.640	237.941	308.066	400.448
39	47.886	59.402	74.401	94.026	119.800	153.762	198.635	258.057	336.882	441.593
40	49.375	61.610	77.663	98.827	126.840	164.048	213.610	279.781	368.292	486.852
EDR*	−0.99	−1.98	−2.91	−3.85	−4.76	−5.66	−6.54	−7.41	−8.26	−9.09

5.9 Appendix B to Chapter 5

Summary of discounting equations

1. Present value of €x received in time t at discount rate r%

$$\text{PV of } x = \frac{€x}{(1 + r)^t}$$

2. Present value factor

$$\text{PVF} = \frac{1}{(1 + r)^t}$$

3. Present value of costs C_t at discount rate r% and project life N years

$$\text{PV} = \sum_{t=0}^{N} \frac{C_t}{(1 + r)^t}$$

4. Present value of annuity of €1 for N years at discount rate r%

$$\text{PVA} = \frac{(1 + r)^N - 1}{(1 + r)^N \times r}$$

5   Annual equivalent (AE) which has present value of costs equal to PV

$$PV = \sum_{t=1}^{N} \frac{AE}{(1+r)^t} = AE \frac{(1+r)^N - 1}{(1+r)^N \times r} \qquad AE = PV \frac{(1+r)^N \times r}{(1+r)^N - 1}$$

6   Present value of a current cost €C with inflation rate $i\%$ and nominal discount rate $r\%$.

$$PV = \sum_{t=0}^{N} \frac{C \times (1+i)^t}{(1+r)^t} = C \times \frac{\left(\dfrac{1+i}{1+r}\right)^N - 1 \times \dfrac{1+i}{1+r}}{\dfrac{1+i}{1+r} - 1}$$

7   Discount rate applied to a cost escalating at $i\%$, with nominal discount rate $d\%$ (d greater than i)

$$r = \frac{1+d}{1+i} - 1$$

8   Net inflation rate for a cost inflating at $i\%$ with nominal discount rate $d\%$ (d less than i)

$$e = \frac{1+i}{1+d} - 1$$

6 Data sources for whole life appraisal

Facilities managers, asset managers, estate directors, project managers and maintenance managers are all involved in managing facilities/assets, and routinely handle running, operating and maintenance information.

6.1 The importance of improving data accuracy

There are six fundamental requirements in the implementation of whole life appraisal:

(1) *A system* by which the techniques can be used: a set of rules and procedures.
(2) *Data* for the proposed project under consideration: estimates of initial and running costs, of performance requirements, service life, design life, discount rates, inflation indices, periods of use/occupancy, energy consumption, cleaning, and so on.
(3) *Professional skill and judgement.*
(4) *Relevant data* that include information on performance and costs of facilities in use measured over time, this could include manufacturers' data on the performance of their products.
(5) *Assumptions made.*
(6) *A feedback system* that provides data about facilities in use.

While the system by which data are analysed is crucially important any system can only provide results that are as good as the original data will allow. If the basic data are inaccurate, then no amount of modelling or sophisticated analysis will give results that are anything other than inaccurate. Effective decision making based on whole life calculations is possible only if the data input to the analytical process is sufficiently reliable to impose reasonable limits on the uncértainty inevitably associated with all forecasting. It follows that there are significant benefits to be gained in the accuracy, reliability, and value of results, from efforts aimed at improving the quality of the basic information and data to which whole life techniques are applied.

Data used in whole life appraisal may be physical, perform-ance, qualitative, or cost data (see Chapter 3). Whole life appraisal has to balance the performance against cost/value; hence this chapter considers possible sources of available cost data. The most important of these data sources – historical data – are examined in some detail.

6.2   Shared data

With the use of spreadsheets, databases and internet-enabled web-based systems, the concept of a shared data environment has become a reality. Data are captured once and used many times. The ability to share data is important. Lessons can be learned from other industries about how data can be accessed in a controlled way. The defence sector uses integrated logistic support (ILS), which will be discussed later.

Data are not costless to collect. It is important to remove the duplication of effort that exists when everybody collects their own data and does not share it with others. It must be consistent, reliable, and reflect the operating environment from which it was collected.

Be aware

- ❑   Avoid 'data dredging', collecting data for data's sake.
- ❑   Ensure there is a feedback mechanism into design, so that the design team can base their decisions on information about costs in use.
- ❑   The UK Office of Government Commerce (OGC, 2002) states: 'A great deal of time can be spent going through lots of historical data from numerous sources in an attempt to get the most accurate information. This process is time consum-ing and normally shows that there are enormous gaps in the data available for creating whole life cost models....It is

always preferable to estimate costs from first principles and to use historical data as a check.'

Be prepared

- ❏ Data are often missing.
- ❏ Data can often be inaccurate.
- ❏ People often believe they have more data than actually exist.
- ❏ It can be difficult to download data for subsequent analysis and for data sharing by a third party.
- ❏ There will be huge variation in the data, sometimes for the same item.
- ❏ Data are often not up to date.
- ❏ Data input is unreliable; the input should be undertaken by those with a vested interest in getting it right.

6.3   Computer aided design (CAD)

Many projects are designed using computer aided design (CAD). When the as built drawings are given to the client/owner, there is frequently an electronic file, the digital equivalent of the paper based technical drawings and manuals. The electronic format makes it possible to provide links to other publications and systems. Interactive electronic technical publications make it possible to link groups of data together in a meaningful way. For example, the maintenance engineer can see in three-dimensional form how a piece of equipment fits into the project. The engineer can rotate the component to see how it can be removed and replaced.

6.4   Basic data sources

Data for whole life purposes are available from three main sources: specialist manufacturers, suppliers and contractors; modelling techniques; historical data.

Data from specialist manufacturers, suppliers and contractors

Most components are bought from specialist manufacturers and suppliers, the construction process being a complex assembly operation on site. Specialist manufacturers and suppliers can be expected to have detailed knowledge of the performance characteristics of their materials and components, but this knowledge will have been influenced by the ways in which facilities are used. Nevertheless, the extensive knowledge and experience of specialist manufacturers and suppliers are a valuable source of whole life information. Wherever possible, manufacturers and suppliers should be asked to provide details of the performance characteristics of the products they are supplying, related to such aspects as expected service life, maintenance requirements, energy use, cleaning requirements, etc.

Specialist contractors should also be consulted. For example, specialist cleaning contractors are able to provide information at the design stage on the cleaning cost implications of particular design decisions and choices of materials. Furthermore, the heavy concentration on repair and maintenance in the construction industry is such that extensive expertise is now available on maintenance requirements and failure characteristics of different components.

It might be argued that the data provided by specialists will be little more than 'best guesses'. To an extent this may be true, but experience and learning should not be underrated. With experience, suppliers, manufacturers and contractors develop a closer understanding of the elements with which they are dealing. As more clients demand this type of information, and subsequently check the information against performance, substantial improvements in accuracy and reliability should ensue.

Five steps are required for the collection of data:

Step 1   Plan the development of the database to hold relevant information on maintenance, cleaning, energy, etc.

Step 2   Plan a monitoring system to collect the data; decide upon who will collect what, and who will be responsible for input.

Step 3   Collect the data.

Step 4   Enter the data into the database.

Step 5   Analysis.

Data from model building

The essence of a model is to reduce a complex system to its essential component parts. Models can be very complex. In weather forecasting, for example, models are used to represent complete meteorological systems. In economic forecasting, the Treasury model of the UK economy consists of more than 500 equations. Nevertheless, these models only approximate reality – a full representation of the UK economy would require almost a full set of equations for every person in the economy. The crucial property of a 'good' model is that it captures the fundamental properties of complicated systems without having to resort to complete representation of those systems.

So it is with models applied to facilities. Particular features of facilities lend themselves to modelling. The models can then be used to analyse the whole life implications of particular design decisions or choices of materials. As an example, the majority of cleaning costs in a commercial building will be incurred in cleaning walls, floors and windows. A simple model could be constructed as follows.

(1) Identify type, location and superficial area of different types of surface finishes such as tiles, carpets, paintwork, glass, etc.
(2) Draw up a flow diagram using the information in item 1.
(3) Estimate labour, material, plant and overhead requirements and costs per unit area of materials to be cleaned, making due allowance for wastage of materials and the sequence of work. These requirements will, of course, vary with the frequency of cleaning and standard of cleaning required.
(4) Apply the unit costs in item 3 to the areas in item 1 using the flow diagram in item 2 to give an overall cost for cleaning related to a cleaning programme.

Clearly much more complicated models can be constructed, but are beyond the scope of this book. What should be clear, however, is that model building is important in whole life appraisal. It can be used, in particular, where historical data are either not available, or are available in too aggregate a form to be applicable and usable, or where the model builder has a better overview of the facility as a system than any other specialist.

Historical data

Historical data sources will produce profiles of initial costs and running costs that are likely to be very different even for the same type of facility. If an elemental breakdown of initial building costs and a similar breakdown of running costs are analysed into the same categories for a proposed facility and compared, an indication will be obtained of those areas in which a whole life cost approach is likely to prove most effective in reducing the total building costs. If a particular building element, e.g. foundations, takes a relatively high proportion of initial costs, but has very low subsequent running costs, little is to be gained from the application of whole life techniques to this element. On the other hand, significant running costs may be indicative of potential whole life cost savings through a change in design decisions. It is then necessary to investigate the sensitivity of the running costs to changes in design or operating procedures.

Figure 6.1 illustrates such comparative cost data for a commercial building. The initial costs are the construction costs. All running cost data have been discounted to a net present cost over a life of 30 years at a discount rate of 5%. Energy costs include all energy-consuming elements, i.e. heating, cooling, ventilation, air conditioning, lighting, hot water and other miscellaneous loads. Maintenance costs include the costs of regular inspection and repair, annual maintenance, and salaries of staff performing maintenance tasks. Replacement of items of less than €4000 in value or having a service life of less than five years are included as part of maintenance. Replacement costs include the cost of replacing the many equipment or other elements with an estimated life cycle shorter than that planned for the entire building.

There are marked differences in the cost breakdowns of these four cost areas. Substructure and superstructure contribute a significant proportion (35%) of initial costs, but account for less than 12% of subsequent energy, maintenance and replacement costs. In contrast, internal finishings and mechanical and electrical services account for a much higher proportion of running costs than of initial capital costs.

The clear implication of Figure 6.1 is that the benefits of a whole life approach are not likely to be spread evenly throughout all elements of a facility. In the first instance at least, attention should be focused on elements such as services and general finishes. It is these elements whose running costs are likely to prove most sensitive to changes in design decisions. As the application of whole life appraisal techniques is reasonably

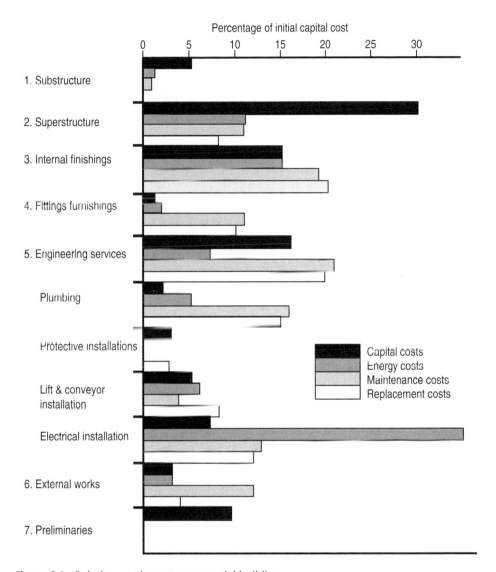

Figure 6.1 Relative cost impact: commercial building.

complicated and time consuming, such effort should be concentrated on the most fruitful areas of a facility.

6.5   Running costs of different types of buildings

The cost analyses discussed previously relate solely to a commercial building and give no indication of the variances of the various averages presented in Figure 6.1. It is to be expected that both initial and running costs will differ quite significantly with facility type, and that these costs will exhibit quite high degrees of variability.

Consider the first of these factors – differences in running costs by building type. Figures 6.2a and 6.2b show the energy used in US residential and commercial buildings. While energy costs are especially important for all building types, other cost areas, in particular cleaning, local taxation and maintenance are also important and on occasions account for a greater proportion of total running costs. The main source of such data comes from the USA where there is greater experience of the application of whole life techniques. It is true that US data cannot be directly transferred to the European environment when making absolute cost comparisons. On the other hand, it is legitimate to use US data to gain an appreciation of variability in costs.

The BMI publish occupancy cost analyses for buildings broken down into occupancy cost elements:

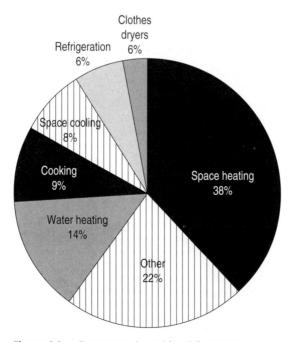

Figure 6.2a   Energy use in residential property.

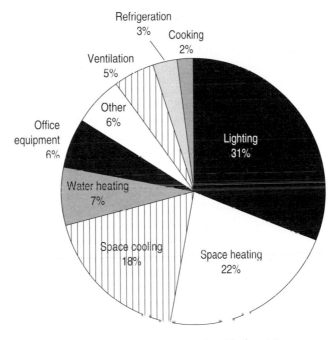

Figure 6.2b Energy use in commercial and industrial property.

(0) Improvements and adaptations
(1) Decoration
    1.1   External decoration
    1.2   Internal decoration
(2) Fabric
    2.1   External walls
    2.2   Roofs
    2.3   Other structural items
    2.4   Fittings and fixtures
    2.5   Internal finishes
(3) Services
    3.1   Plumbing and internal drainage
    3.2   Heating and ventilating
    3.3   Lifts and escalators
    3.4   Electric power and lighting
    3.5   Other M&E services
(4) Cleaning
    4.1   Windows
    4.2   External surfaces
    4.3   Internal

(5) Utilities
    5.1 Gas
    5.2 Electricity
    5.3 Fuel oil
    5.4 Solid fuel
    5.5 Water rates
    5.6 Effluents and drainage charges

(6) Administration
    6.1 Services attendants
    6.2 Laundry
    6.3 Porterage
    6.4 Security
    6.5 Rubbish disposal
    6.6 Property management

(7) Overheads
    7.1 Property insurance
    7.2 Rates

(8) External works
    8.1 Repairs and decoration
    8.2 External services
    8.3 Cleaning
    8.4 Gardening

The sample sizes are quite small for certain facility types, so reducing the confidence that can be placed in the estimates. A number of factors need to be taken into account in using the estimates such as:

- building size, shape and layout;
- specification – levels of insulation, type of glazing, building management systems, maintenance levels;
- intensity of use and performance characteristics – hours, level and type of use; and
- location and orientation – large north-facing areas require more heating.

The BMI's comprehensive collection of occupancy cost information includes an annual maintenance price book, quarterly cost briefings and special reports. Figure 6.3 shows the typical annual expenditure for each of the cost elements for different types of facility. The wide difference in average running costs is worthy of note.

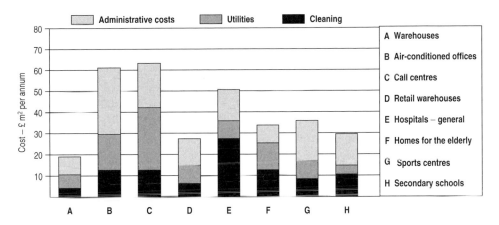

Figure 6.3 Annual occupancy costs by building type.
Source: *Review of occupancy costs 2002*, BMI Special Report 311.

International operating costs

There are a number of studies on operating costs around the world. Multinational clients and owners are interested to know how cities compare. The comparison is not easy because differences in exchange rate, for example sterling moved +20% against the US dollar in 2004. Also, the basis of comparison will always be different; energy costs include air conditioning in Houston, Texas, whereas Reykjavik in Iceland will have very high levels of insulation to minimise heating costs for most of the year. Similarly, a comparison between north and south China will be very different because of the regulations governing heating and cooling and the climate.

Most significant occupancy costs are labour and energy, with labour accounting for about 60% of the total costs, and energy making up a further 20%. The level of labour employed in building maintenance is substantial and as labour rates vary widely across the world, this has a significant impact on the occupancy costs. The E. C. Harris Global Review (2002) gives a guide to air-conditioned office occupancy costs based on countries. Costs need to be used in the knowledge that energy and labour costs will vary not only between countries, but also within countries.

The variability of running costs

Running costs will be affected by:

❑ Different standards of construction.
❑ Age of the facility – as facilities age, they need more attention and renewal.
❑ Climate.
❑ Different occupancy patterns and use and abuse.
❑ Maintenance standards and quality expectations.
❑ Size of the facility – affecting economies of scale and buying power.
❑ New legislation that requires a facility to be changed/upgraded in order to comply.

The variations in the primary cost elements used by the BMI can include:

❑ Cleaning – the need for extra hygiene in food preparation areas, hospitals, health centres will place a premium on the cleaning costs.
❑ Utilities – the size, operational hours and temperature requirements of a facility will affect the cost of utilities.
❑ Administrative costs – these are dependent on the management of the facility and so there can be large variations in the cost, even if the facilities are similar.
❑ Overheads – made up of local taxation and insurances and is dependent on many factors, the most important of which is location.

Because of these variations, data have to be selected for analysis in a very structured way. Aspect, user function, types of external cladding, and many other features all carry implications for running costs. Strong statistical techniques can be applied only if data are stratified according to such cost-significant parameters.

Some cautionary words

A standard set of definitions and approach to capturing data is essential, otherwise difficulties of comparison will persist. Time can be wasted on sterile arguments about the meaning of figures and it will be difficult to link operating cost and performance data with high-level business data, which is necessary for the effective management of any business.

Cost indices

Table 6.1 shows a comparison of various indices over a number of years.

Table 6.1 Occupancy costs indices.

	1993	1994	1995	1996	1997	1998	1999	2000	2001
Cleaning	122.2	124.1	127.3	131.1	138.2	143.2	146.5	150.6	157.7
Fuel oil	85.3	96.8	110.5	119.3	112.5	95.0	103.3	139.5	135.3
Electricity – industrial	105.3	101.3	99.5	96.3	90.8	90.0	90.5	84.0	81.5
Gas – industrial	99.0	100.0	87.3	63.8	65.8	69.5	68.8	74.5	105.0
BMI Energy Index	110.3	112.8	113.0	110.8	109.8	106.8	104.8	107.0	111.0
Retail Price Index	140.7	144.1	149.1	152.7	157.5	162.9	175.5	180.5	183.5
BCIS General Building Cost Index	149.8	155.0	162.0	167.5	171.8	178.3	182.8	190.5	196.0
Average Earnings Index	93.6	97.0	100.0	103.6	108.0	113.5	119.0	124.3	129.8

Source: *Review of occupancy costs 2003*, BMI Special Report 322 (April 2003).

There is a sharp contrast between costs of air-conditioned and non-air-conditioned offices. Some of the differences are more apparent than real. There is a difference of anything up to a factor of four in heating and electrical costs between air-conditioned and non-air-conditioned buildings. This arises because the operation and maintenance of air-conditioning systems demand more operations staff in addition to any extra energy demands they impose.

Two important conclusions can be drawn from the variability of running costs. First, it would appear that running costs are not 'given' for any selected facility type. Significant variation remains in exhibited performance, at least part of which is rooted in design decisions. In other words, there is potential for reduction in running costs and total whole life costs by the application of whole life cost techniques.

Second, work remains to be done in identifying the cost significant parameters by which historical data should be stratified for whole life cost purposes. Facilities of 'similar' types should be selected for cost estimation.

There remains, of course, the question of how historical data might be obtained. The text has concentrated upon published data and shown that these sources, while useful, are not of themselves adequate for whole life appraisal purposes. Nevertheless, just as experience in the application of whole life cost will improve data drawn from specialist suppliers and model building, so experience and increased use of sources such as BMI can be expected to improve the usefulness of these data.

6.6   Data on durability and life expectancy of materials

A number of research organisations provide guidance on durability. The Housing Association Property Mutual (HAPM) (an organisation that provides insurance for social housing) *Component Life Manual* schedules insured lives and maintenance requirements for over 500 domestic building components. Table 6.2 shows example data from the manual. The lifespan assessment takes into account the component material and the quality of assembly and protection, which can be adapted for life-cycle costings or planning maintenance programmes. The manual is divided into seven sections each representing component groups.

Table 6.2   Example data from the HAPM *Component Life Manual.*

Doors, windows and joinery – hardwood			
External masonry walls	External walls timber-framed	Description	Maintenance
B1	B1	Permeable hardwood (e.g. Opepe, English elm) double vacuum impregnated with organic solvent preservative to BWPA schedule V1 or V2 or non permeable (e.g. Idigbo, Keruing, Luan, Mahogany, Meranti, European Oak, Utile) to schedule V3, after machining. Joints combed or morticed, held with plated/non-ferrous mechanical connection aids. Joints fully coated in 'WBP' adhesive so as to seal end grain. Timber quality at least 'Class J50' frames, 'Class J30' sashes, 'Class J40' beads (BS EN 942). PSA approved timber window ranges (MOB 08 specification).	Stain at 3 years, paint at 5 years. Lubricate ironmongery, replace glazing compounds, weatherstripping, etc. as required. Replace unpainted ironmongery (e.g. scissor hinges/espagnolettes) plated at 10 years or stainless steel at 20 years.

Doors, windows and joinery – hardwood			
External masonry walls	External walls timber-framed	Description	Maintenance
B2	B2	Heartwood only of untreated hardwood of a species designated as 'SW' or 'SWC' (suitable) for external use without preservation in Table NA2 of BS EN 942 (e.g. Afrormosia, Iroko, Mahogany, [but not 'Philippine' mahogany], European Oak, Sapele, Teak Utile). Joints combed or morticed, held with plated/non-ferrous mechanical connection aids. Joints fully coated in 'WBP' quality adhesive so as to seal end grain.	Stain at 3 years, paint at 5 years. Lubricate ironmongery, replace glazing compounds, weatherstripping, etc. as required. Replace unpainted ironmongery (e.g. scissor hinges/espagnolettes) plated at 10 years or stainless steel at 20 years.
B3	B3	Permeable hardwood (e.g. Opepe, English elm) double vacuum impregnated with organic solvent preservative to BWPA schedule V1 or V2 or non-permeable (e.g. Idigbo, Keruing, Luan, Mahogany, Meranti, European Oak, Utile) to schedule V3, after machining. Joints combed or morticed, held with plated/non-ferrous mechanical connection aids. Joints fully coated in type 'D4' adhesive to BS EN 204 so as to seal end grain. Timber quality at least 'Class J50' frames, 'Class J40' sashes, 'Class J30' beads (BS EN 942).	Stain at 3 years, paint at 5 years. Lubricate ironmongery, replace glazing compounds, weatherstripping, etc. as required. Replace unpainted ironmongery (e.g. scissor hinges/espagnolettes) plated at 10 years or stainless steel at 20 years.
C1	C1	Permeable hardwood double vacuum impregnated with organic solvent preservative to BWPA schedule V1 or V2 or non-permeable species to schedule V3, after machining. Joints combed or morticed, held with plated/non-ferrous mechanical connection aids. Joints fully coated in 'MR' adhesive or type 'D3' adhesive to BS EN 204 so as to seal end grain.	Stain at 3 years, paint at 5 years. Lubricate ironmongery, replace glazing compounds, weatherstripping, etc. as required. Replace unpainted ironmongery (e.g. scissor hinges/espagnolettes) plated at 10 years or stainless steel at 20 years.

(*Continued*)

Doors, windows and joinery – hardwood			
External masonry walls	External walls timber-framed	Description	Maintenance
E1	E1	Non-permeable hardwoods, or 'mixed' species, double vacuum impregnated with an organic solvent preservative to BWPA schedule V1 or V2 after machining. Joints combed or morticed, held with plated/non-ferrous mechanical connection aids. Joints fully coated in 'MR' quality adhesive or type 'D3' adhesive to BS EN 204 so as to seal end grain.	Stain at 3 years, paint at 5 years. Lubricate ironmongery, replace glazing compounds, weatherstripping, etc. as required. Replace unpainted ironmongery (e.g. scissor hinges/espagnolettes) plated at 10 years or stainless steel at 20 years.
E2	E2	Permeable hardwood, treated with a 3-minute immersion in an organic solvent preservative after machining. Joints combed or morticed, held with plated/non-ferrous mechanical connection aids. Joints fully coated in 'MR' quality adhesive or type 'D3' adhesive to BS EN 204 so as to seal end grain.	Stain at 3 years, paint at 5 years. Lubricate ironmongery, replace glazing compounds, weatherstripping, etc. as required. Replace unpainted ironmongery (e.g. scissor hinges/espagnolettes) plated at 10 years or stainless steel at 20 years.
F1	F1	Non-permeable or mixed species dipped/immersed in organic solvent for minimum 3 minutes. Joints combed or morticed, held with plated/non-ferrous mechanical connection aids. Joints fully coated in 'MR' quality adhesive or type 'D3' adhesive to BS EN 204 so as to seal end grain.	Stain at 3 years, paint at 5 years. Lubricate ironmongery, replace glazing compounds, weatherstripping, etc. as required. Replace unpainted ironmongery (e.g. scissor hinges/espagnolettes) plated at 10 years or stainless steel at 20 years.
G1	G1	Untreated hardwood of a species designated as 'SP' or 'SPC' (only suitable for external use if preservative treated) in Table NA2 of BS EN 942 e.g. Luan, Meranti, Agba).	Stain at 3 years, paint at 5 years. Lubricate ironmongery, replace glazing compounds, weatherstripping, etc. as required. Replace unpainted ironmongery (e.g. scissor hinges/espagnolettes) plated at 10 years or stainless steel at 20 years.

Doors, windows and joinery – hardwood			
External masonry walls	**External walls timber-framed**	**Description**	**Maintenance**
U1	U1	Untreated hardwood of indeterminate species or of a species designated as 'X' (unsuitable) for external use in Table NA2 of BS EN 942 (e.g. Ash, Beech, Ramin, Sycamore) and/or timber quality, adhesive, or mechanical fixings inadequate.	None

Source: HAPM web page www.hapm.co.uk

According to BS 7543:1992, the main causes of deterioration in facilites are:

❑ action of weathering;
❑ biological infestation;
❑ stress;
❑ chemical interactions;
❑ physical interactions; and
❑ wear and tear.

This deterioration can be accelerated by a number of different factors that result from human actions and so are difficult to predict. The factors are:

❑ poor design/detailing;
❑ inappropriate selection of material or component for intended use;
❑ quality of material or component used;
❑ adverse on-site storage and handling;
❑ poor workmanship;
❑ inadequate maintenance; and
❑ inappropriate use.

The processes of design and construction are aimed at minimising the factors, but it is often a combination of them that leads to failure (BSI, 1992)

The life expectancy of a material, element or component does not just involve its physical life. Many components are replaced because of obsolescence – see page 41. A BMI survey (BMI, 2001) considered three levels of component life: the minimum life, the maximum life and the typical life expectancy. Some results of the survey for a selection of components are shown in Tables 6.3, 6.4 and 6.5.

Table 6.3 The life expectancies of render to blockwork wall.

Average typical life	Average minimum life	Average maximum life
❏ 55 years	❏ 35 years	❏ 80 years
Factors to be considered when assessing the life expectancy		
❏ Exposure ❏ Local air quality ❏ Orientation	❏ Thermal and moisture movement ❏ Block type and quality ❏ Position (parapet, under window, wall, etc.)	❏ Mortar joint quality and strength ❏ Render specification ❏ Chemical interaction between render and mortar
Early deterioration may be attributed to the following		
❏ Wall movement ❏ Atmospheric pollution ❏ Impacts	❏ Differential movement between render and blockwork ❏ Simple loss of adhesion between render and wall ❏ Excessive loading	❏ Differential shrinking between 1st and 2nd coats of render ❏ High level of sulphates in block mortar joints ❏ Continual wetting of render

Source: *Life expectancy of building components*, BMI, 2001.

Table 6.4 The life expectancies of bitumen felt covering to flat roof.

Average typical life	Average minimum life	Average maximum life
❏ 20 years	❏ 10 years	❏ 25 years
Factors to be considered when assessing the life expectancy		
❏ Exposure and roof size ❏ Local air quality	❏ Full or partial bonding workmanship ❏ Inspection/ maintenance regime	❏ Inappropriate detailing ❏ Adequate solar protection
Early deterioration may be attributed to the following		
❏ Deck movement ❏ Foot traffic	❏ Moisture or air between layers ❏ UV degradation	❏ Thermal movement ❏ Atmospheric pollution

Source: *Life expectancy of building components*, BMI, 2001.

Table 6.5 The life expectancies of a boiler.

Average typical life	Average minimum life	Average maximum life
❏ 20 years	❏ 15 years	❏ 30 years
Factors to be considered when assessing the life expectancy		
❏ Boiler material ❏ Fuel type ❏ Adequate ventilation ❏ Water hardness ❏ Blocked pipes	❏ Material quality ❏ Vented/unvented system ❏ Boiler efficiency ❏ Operating pressure ❏ Water treatment	❏ Boiler type ❏ System design ❏ Required output ❏ Operating temperature ❏ Maintenance regime
Early deterioration may be attributed to the following		

❏ Improper operation	❏ Inappropriate
	   installation
		❏ Failure of controls,
		   valves, fan and pumps

Source: *Life expectancy of building components*, BMI, 2001.

7 Operations and maintenance

This chapter is divided into two main sections:

❑ Maintenance
❑ Energy in buildings

7.1 Maintenance

This section suggests an approach to the consideration of maintenance for whole life purposes. It does not deal specifically with the design factors that cause maintenance problems, nor with the remedial aspects of defects.

One of the most difficult decisions facing facility/asset owners is the timing of different types of maintenance work in order to keep facilities up to a proper and acceptable state of repair. Several policies are available, ranging from a short-term temporary repair to the undertaking of a full-scale renewal. The choice of action depends on a number of factors, and, as a result, significant reductions in maintenance costs are likely to result from a whole life approach:

❑ The rate of deterioration.
❑ The cost of different types of repair.
❑ The disruption and disturbance to the building occupants and time required for the repair.
❑ The relationship between the physical life of the repair and the required physical, functional, and economic life of the facility.

Inevitably, facilities wear out, require maintenance and repair, and must eventually be replaced, because materials have a finite life. Not all of the elements of a facility will deteriorate at the same rate. Some, such as the foundations, will last as long as the facility and will require minimal corrective expenditure, while others, such as exterior paintwork, will deteriorate and will need frequent attention. The need for maintenance work also arises because of excessive or abusive use, vandalism, faulty design, bad workmanship, poor quality material, or inaccurate specification. In addition, the appropriate maintenance programme and targets will vary with the quality of the overall facility, for example, a bank will require very different maintenance from a factory store.

In theory, maintenance should be constantly reviewed and the fabric and services kept to an acceptable state of repair. Precise definition of an 'acceptable state of repair' is difficult and will vary from element to element. There is a substantial difference between maintenance of plumbing, mechanical and electrical services, and maintenance of the fabric. When a heating system fails, it will generally require immediate attention if the facility is to remain operational. The fabric, on the other hand, does not usually fail catastrophically and the facility can continue to be used even if the fabric has significantly deteriorated.

Maintenance repair consists of three stages: inspection, diagnosis and constructional or remedial action. For whole life purposes, the maintenance can be classified under the following subheading.

Maintenance of the main structure

The main structure is exposed primarily to the natural elements and the maintenance work will probably involve inspection and routine planned maintenance. Whole life appraisal can be used to determine the best form for this maintenance programme.

Maintenance of the finishings/fixtures/fittings

The finishings suffer from wear and tear by the occupants and will require periodic renewal. Again, whole life appraisal can be used, for instance to determine optimal intervals for renewal.

Modernisation and adaptation

This will often take place on a planned basis at a certain point in the facility's life.

Maintenance of the external works

The external works will require extensive maintenance with grass cutting, replacement of shrubs and trees, and paving.

Maintenance of the plumbing, mechanical and electrical services

Each of the service elements will have its own maintenance requirements. While planned and preventive work will be undertaken, frequent corrective maintenance dependent upon the age will be needed.

Redecorations

Internal and external redecorations will be necessary on a planned basis. Only in exceptional circumstances will the complete redecoration of the facility take place; the more likely course is partial redecoration on a rotating basis, the redecorating programme being determined by a whole life cost approach.

Maintenance to bring to an acceptable standard to comply with new legislation

While the application of whole life appraisal techniques to maintenance costs will generate cost savings, a number of problems will arise in their implementation.

❑ Maintenance cost data are difficult to classify to a detailed level because of the varied nature of the work.
❑ Reliable information on deterioration rates is not readily available.
❑ The estimated repair and replacement costs are highly variable. For example, when restoring internal paintwork to as

good as new condition, the cost will vary according to the age of the element. If the internal work is five years old it will be sufficient to wash down and paint with two coats of paint, whereas if the paintwork is ten years old it will probably need stripping back, filling, and painting.

❏ The deterioration rates can represent only an average. Elements may deteriorate in the same general pattern, but because all buildings are designed and used differently, there will be a corresponding difference in the replacement cycle. However, the expected renewal times are required for planning purposes.

❏ People's perception of an acceptable standard or quality will be different.

❏ There is no standard methodology for cost planning maintenance work.

Several methods can be suggested for overcoming these problems. First, historical data are available. Second, specialist materials and components' suppliers should be asked to provide performance characteristics for their products. In particular, they should be able to provide reasonably accurate information on probable system life and required maintenance.

Finally, a model building approach can be adopted. This is best described as a series of steps.

Model building approach to maintenance

Step 1

List the maintenance elements.

Step 2

Decide upon annual maintenance and intermittent maintenance. An example of an intermittent schedule is shown in Table 7.1.

Step 3

Apportion a unit of measurement together with the quantity for each elemental category. Some caution is advised, however: if 50% of the carpet is replaced in year 10, it does not necessarily mean

Table 7.1 Intermittent maintenance schedule.

Element	Service life before replacement (years)	Proportion of failure	Replacement cost	Repairs and maintenance		
				Repair	Interval	Cost
Heating						
Boiler pump – direct drive pump	15	100%	€190			
Gas fired boiler	20	100%	€1600			
Boiler flue 250 mm diameter	20	100%	€160			
Control equipment	25	100%	€410			
Floors						
Vinyl tiles	15	75%	€18/m^2	Patching	5 years	€6/m^2
Carpet	10	50%	€22/m^2	Patching	5 years	€8/m^2
Woodblock	60	100%	€40/m^2	Re-sanding	10 years	€2/m^2
Quarry tiles	60	100%	€28/m^2	Re-seating	10 years	€4/m^2

that the remaining 50% will last a further 10 years. Hence, a spreadsheet should be used to plan how the carpet will be replaced over the whole time horizon. If a 35-year time horizon is being used, the boiler would be replaced in year 15 and year 30, thereby leaving 10 years at the end of the time horizon. Reality must prevail and the likelihood with such an expensive item is that it will be refurbished rather than replaced. These types of maintenance plans must be undertaken, particularly for PFI/BOT projects.

Step 4

Consider each of the elements with their likely condition over the lifespan of the facility or the time horizon of the analysis. Table 7.2 shows the age and condition for the external doors and windows element. Five-year time periods and four condition bands have been used in the example. A choice must be made about whatever time period is felt appropriate. The figures have been chosen purely for illustrative purposes. The table is showing a 'do nothing' situation, whereas it is likely some work will have been undertaken. By applying this procedure systematically in all the time periods, it is possible to build up a comprehensive

Table 7.2 Age and condition for the external doors and windows elements.

Condition	Age (years)					
	0	5	10	15	20	25
1	100	80	60	40		
2		20	20	30	40	
3			20	25	30	60
4				5	30	40
	100%	100%	100%	100%	100%	100%

Condition 1 – as good as new
Condition 2 – generally acceptable condition with minor modifications
Condition 3 – requires major modifications
Condition 4 – unacceptable condition

picture of the deterioration of each element over time. (As was stated above, the deterioration rates are averages.)

Step 5

The table shows the deterioration if no maintenance work is undertaken. Allocate present day prices to the tasks involved to identify the optimum expenditure on maintenance. The table also highlights two aspects of maintenance. First, it gives a checklist of items to be considered within the maintenance category. Second, the condition bands put a time value to the deterioration rate. The chart should, of course, be treated as a first stage in the analysis of maintenance costs using a structured model. It is, however, capable of extensive development by those professionals in the industry with the precise information and the expert skill and judgement necessary for such development.

There is no magical solution to maintenance planning, nor to the ways in which such planning can be implemented in practice. Planning is essential if facilities and elements are to be chosen and operated in a cost effective way. The data required for maintenance cost planning will not always be easy to obtain. Three techniques have been suggested by which such data can be generated. The technique to be used in any given situation will be dependent upon the precise circumstances – a model building approach has certain attractions but must be supported by the collection and analysis of historical data. Experience in the analysis of maintenance costs will lead to significant improvement in the quality of both data and techniques.

7.2   Energy in buildings

Energy conservation has become critical in the planning and design of facilities owing to increasing energy prices and the threat of fuel shortages. Supplying energy will become increasingly expensive. Maintaining the current levels of demand for energy in facilities involves unprecedented economic risk due to the instability of the world energy market. Energy consumption in building systems accounts for approximately 45% of the UK primary energy demand. Figure 7.1 shows the breakdown of energy consumption within a commercial building during its operational phase, over a 50-year life. As would be expected, the highest percentage of total consumption was for usage – including space and water heating, and cooling.

Whole life planning can be used to select energy conservation measures both at the design stage and in the retrofitting of existing facilities. Its emphasis is on determining how to allocate a given budget among competing options so as to maximise the net return, the main aim being to save energy in ways that are cost effective. This is reinforced by the Department of Energy Circular 1/75, which states: 'It is Government policy that energy conservation measures taken should be fully justified in economic terms, their cost being completely covered by the fuel saving achieved so that there is no waste of resources as a whole'.

About 75% of the building stock in the UK, 35% of which was built before 1945, is likely to continue in use to the end of the twenty-first century and beyond. Most of these facilities were

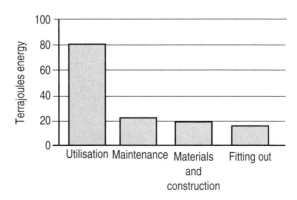

Figure 7.1 Energy consumption in commercial building over its 50-year life. Source: International Aluminium Institute.

designed at a time when fuel costs were much lower and the financial incentive to reduce energy consumption was much less. Each year new buildings add less than 1% to the existing building stock, and thus the main opportunity for energy conservation lies in retrofitting the existing stock of buildings. Comparing the total costs of energy conservation projects is not straightforward because of the interaction between the running cost elements. Costs, which must be considered over the whole life in addition to energy costs, are repair, replacement, maintenance and operating personnel, such as engineers. The fuel costs are the costs of delivered energy to the building, but these are influenced by the operations and maintenance policy of the building. More expenditure on planned maintenance will improve the efficiency of equipment and reduce its energy usage. Furthermore, insurances and local taxation will be influenced by an improvement in certain facilities in the building.

Whole life planning and energy conservation

This section will give a broad outline of those factors to be considered when preparing a whole life cost plan for energy conservation projects. The simplest approach is to consider a series of steps.

Step 1 Identify the various options.

Step 2 Establish the appropriate life cycle and discount rate to be used in the analysis.

Step 3 Identify all running cost elements and likely benefits associated with the options.

Step 4 Estimate the initial capital costs of the options.

Step 5 Estimate the facility's energy requirements in the light of its construction, usage characteristics, and environmental requirements.

Step 6 Estimate the energy costs, the annual operations costs, the annual maintenance and intermittent costs. Discount future costs to present values, if making comparisons of options.

Step 7 Rank the options and test the sensitivity to the various assumptions underlying the estimates. For example, examine how a reduced replacement cycle for the boiler in a heating system affects the whole life cost.

Estimating energy requirements

The calculation of the energy requirements and heating and cooling equipment sizes in a facility is very complex. It not only requires determination of the heating and cooling loads, taking into account the continually varying outside weather and the frequently varying inside load conditions, but also determination of the performance of the mechanical systems under conditions of partial load. The peak capacity requirements of equipment must be identified as well as the translation of the building operating schedules into energy demand and consumption.

The services engineer will normally provide an estimate of the energy requirements for a proposed facility. There are also a considerable number of energy estimation models available, ranging from the highly sophisticated computer assisted models through to simple empirical approaches. Many of the design data used in the models derive from the Chartered Institution of Building Services Engineers (CIBSE) design guides. The model developed in the following paragraphs is intended as a simple approach that could be used at the early design stages of a project. It is not envisaged as a substitute for the expert advice provided by the services engineer.

How and why buildings use energy

One important factor that determines the energy consumption is the way the facility is used. It is the occupants who place the demands on systems that use energy. The reality of such statements is often ignored in many forecasts of energy use. It will be apparent that the hours of operation have a significant impact upon the energy use. A hospital or hotel is used throughout the year for 24 hours a day, whereas a school is used for approximately 7 hours a day for 40 weeks of the year.

Comparisons of the energy consumption of similar functional facility types on the basis of installed capacities of heating, ventilating, air conditioning, lighting and other electrical equipment can be misleading, because the amount of energy used in a facility will be a function of the following:

❑   Temperature and humidity levels.
❑   The number of occupants and the task they perform.

❏ The size, shape, and zoning.
❏ Ambient weather conditions.
❏ The window/wall ratio, the type of glazed areas and the provision of shading.
❏ The 'U' values of the building fabric which affect heat loss, heat gains, and the thermal response.
❏ The lighting levels.
❏ Air movement and ventilation rates.
❏ The hours of operation of systems and components.
❏ The orientation (the effects of winds and solar heat gain in relation to the shape).
❏ The presence of an energy management programme.
❏ Heat losses and heat gains for each building zone.
❏ The different fuels used.
❏ The extent of appliances and machinery.
❏ The presence of heating control systems.
❏ The number of lifts.
❏ The operating efficiencies of the heating and cooling equipment and distribution systems.

Estimating the energy requirements for a facility is as problematical as estimating a construction price for a proposed building at the design stage. Prices are influenced by the marketplace; energy use is influenced by the weather, occupancy patterns, and other variables. As an illustration, theoretically, the temperature in all rooms of a building having the same function should be controlled exactly to the same value. However, in reality the temperature in any one room swings about a mean owing to variations in occupancy, solar gain, and so on. Temperature fluctuations should be kept to a minimum. A rise of 1 °C in mean room temperature from 18 °C to 19 °C is accompanied by a rise of approximately 10% in fuel consumption for space heating.

Sustainable energy use

Energy consumption costs money and has an impact on the environment in terms of use of natural resources and emissions. The growing emphasis on energy both for industry and domestic use has led to systems of measurement and efficiency as well as the search for renewable energies.

Whole life appraisal necessitates an awareness of energy usage and costs and the levels of embodied energy in building materials and components. Over the whole life of a facility legislation and emission standards are likely to change and anyone undertaking a whole life appraisal needs to be aware of this.

Using sustainable energy sources with minimal emissions will reduce the risk of contravening regulations that are likely to become more stringent.

Environmental accountability

Environmental accountability is permeating many aspects of construction worldwide. The 'Nordic Swan' eco-label established in 1989 is the only multinational environmental labelling scheme in use mainly for consumer products. In the construction industry there are few examples of eco-labelling. Legislation is increasing to ensure that everybody involved in the construction process pays greater attention to the long-term implications of buildings and constructed facilities.

Suppliers, manufacturers, members of the design team, contractors and specialist trade contractors are all having to respond to this demand to comply with increasing environmental legislation. Furthermore, environmental commitment is now seen as an essential component of good commercial housekeeping. Our buildings must incorporate energy conservation measures and they should use less environmentally damaging materials and products. Environmental impact assessments are now carried out as a legal planning prerequisite for certain types of applications. These assessments gather information about a project, and the probable effects and impact on the environment, with the regulations providing an assessment of all the potentially significant environmental effects of a proposed development.

Some of the key aspects of environmental management in the built environment include:

❏ Source control (verifying origins of environmentally sensitive materials, for example, chemical treatments).
❏ Embodied energy tracking.
❏ Waste management.
❏ Energy management.

❏   Maintenance management (increasing the longevity of components by condition monitoring).
❏   Waste disposal and reclamation management.

A major long-term concern is the dismantling and disposal of facilities and their components. Legislation demands that records are kept and made available to demonstrate the safe disposal of any waste materials. Waste regulation authorities receive records from waste-disposal operators and have to collate and analyse the data. Under legislation, a complete record of transfer of waste from producer to final disposal must be maintained and kept for at least two years. Operators of waste disposal plants and landfill sites have become very aware of the need to establish the origin of waste that is being used in their industry.

The ISO 14000 series addresses environmental management systems, environmental auditing, environmental labelling, environmental performance evaluation, and life cycle assessment. ISO 14000 requires the development, implementation, and maintenance of environmental management systems aimed at ensuring compliance with stated environmental policy and objectives. The standard, *Guide to Environmental Management Principles, Systems and Supporting Techniques*, does not itself lay down specific environmental performance criteria. It states that

> ...the organisation shall establish procedures for the identification, collection, indexing, filing, storage, maintenance and disposition of environmental management records. Pertinent contractor and procurement records...shall form an element of these records. Records shall be stored and maintained in such a way that they are readily retrievable and protected against damage, deterioration or loss.

The document handling demands resulting from such measures could be punitive to the design and construction processes which are critically affected by process times.

Tracking the data

The attention to environmental issues has inevitably resulted in volumes of data being collated, processed and analysed. The sheer volume of manual processing of paper leads to difficulties of tracking information through the life of the project. ISO 14020 (Environmental Labels) addresses the issue of environmental labelling, but is concerned with the definition and certification of product claims rather than the technological implementation of a labelling system. The use of automatic data capture and tracking (automatic identification) linked to computer systems would appear to give a vehicle for achieving effective and efficient capture, tracking and management of environmental data and so avoid some of the problems resulting from this development. In order to establish an integrated environmental information system, a method of environmental data labelling is essential. A standard method of labelling should:

❑ Minimise the time involved in collecting the initial data from suppliers.
❑ Provide a 'cradle to grave' approach to materials tracking throughout the supply chain.
❑ Improve the accessibility of key environmental data when needed, by numerous stakeholders.
❑ Enforce a degree of rigour with respect to the type of information and its interpretation.
❑ Facilitate the efficient disposal, recovery and reuse of materials at the end of their service life.

8 A practical approach to whole life appraisal for construction

8.1 Introduction

This chapter considers some of the practical aspects of using whole life appraisal when examining the running costs of facilities. In view of some of the problems outlined in the previous chapters, the reader at this stage may be feeling concerned at the difficulties of comparing facilities to evaluate whole life costs and performance. However, in the same way that cost consultants and contractors are able to forecast construction prices at the very early design stages of a proposed project, a similar knowledge will be acquired with the benefit of experience to deal with the whole life and running cost aspects.

8.2 Initial problems

Forecasting

A fundamental aspect of whole life appraisal is that it embodies forecasting, the purpose of which is to provide information for decision making. All organisations make forecasts, although some do not use formal or scientific methods. Cost consultants forecast tender prices, construction enterprises forecast their costs of construction and the time taken to complete the project, specialist contractors forecast the cost of their installations.

The fact remains that there is no infallible way to predict the future; forecasting is an art, not an exact science. It is, nevertheless, an area of decision making that cannot be left to go by default. Forecasting will not guarantee correct decisions, but it will improve the basis on which decisions are made. Thus whole life appraisal will give a clearer vision of the future than could be achieved by intuition alone. The effort will be justified even if it merely leads to the rejection of a few demonstrably wrong decisions.

Where the budgets for capital expenditure, maintenance, and utilities (energy, water) are not vested in the same individual, the holders of the separate budgets will need to work together to arrive at the optimum whole life cost and performance solution.

It is important to focus attention on those elements likely to have the most significant impact upon the running costs. The objective is to ensure that the client obtains value for money from both an initial cost and a total cost approach.

Risk and uncertainty

When considering the total cost approach, it is essential to draw clear distinction between what could happen, what should happen, and what will happen. Obviously, clients are interested in what will happen. However, facilities are used in a variety of ways, and the distinction between a feasible use and what actually occurs involves a complicated network of human decisions, most of which are taken by people who may not have been directly involved in the design process. For example, a facility might be cleaned at night necessitating lighting for the cleaners, whereas the design team may have assumed that cleaning would take place with minimal artificial lighting.

8.3   Using the weighted evaluation technique as a decision-making tool

It is not sufficient to make choices of say the floor covering of a facility based on cost alone. The performance criteria need to be considered. Dell'Isola and Kirk (1981) modified a management technique to develop the weighted evaluation technique. The following example considers a choice of floor finishes. The various criteria are weighted numerically according to the importance that is attached to them and the finishes are scored according to the degree to which they meet the criteria. Finally, an overall score is calculated for each finish.

Each of the criteria that the finish must satisfy are listed (A, B, C, etc.) in the criteria scoring matrix in the top half of the form. Each is then systematically compared with every other criterion in terms of direction and strength of preference. For example, for this particular functional space, 'water absorption' (criterion G) is given a minor preference over 'moisture movement' (criterion D) and so is assigned a score of 2 (G-2) in the appropriate cell of the criteria scoring matrix. Water absorption and slip resistance (criterion E) are ranked equally, giving the entry E/G, while bacterial/microbial resistance (criterion A) is given a major preference over every other criterion – the ranking A-4.

The process is repeated until all the criteria have been compared with one another and the results recorded in the appropriate cells of the matrix. The raw scores for each criterion are found by adding up the numbers in each cell where a preference for a particular criterion has been registered. Take, for example, criterion G, see the shaded cells in Figure 8.1(a). To obtain the raw score for G, all the cells in the two diagonals originating from G are examined. The upward sloping diagonal contains G-3, E/G, G-2, G-3, G-3, A-4 and the downward sloping diagonal G-4, G-3. This gives a total score for G of $3 + 1 + 2 + 3 + 3(+ 0) + 4 + 3 = 19$. Note that the 'no preference' entry E/G scores 1 and A-4 scores zero. This is done for every criterion to generate the raw score line in Figure 8.1(a). This set of numbers identifies the relative importance of the various criteria in this particular functional space. Thus in this example bacterial/microbial resistance is very important, while stain resistance and low cleaning cost are of very minor importance relative to the other criteria.

These raw scores can now be used directly in the next stage of the exercise. However, for convenience, they may be converted into a set of weights on a scale of 1 to 10. To change raw scores R, into weights W, use the formula: $W - 10R/M(n - 1)$, where, M = maximum score for each criterion (4, in the example) N = number of criteria, (9 in the present case). Thus in Figure 8.1(a) $W = 10R/4(9 - 1) = 0.3125R$.

A raw score of 14 for criterion E then gives a weight of $14 \times 0.3125 = 4$, as entered in the weight line of Figure 8.1(b). The advantage of this is that the second stage of the work may be undertaken without a calculator, no matter how large the criteria matrix.

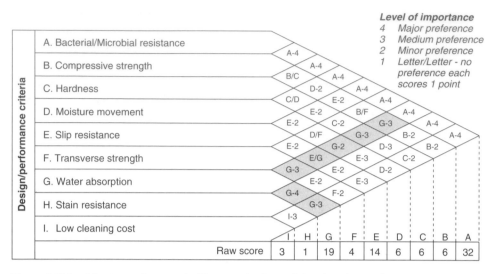

Figure 8.1(a)   Criteria scoring matrix (first step in the weighted evaluation).

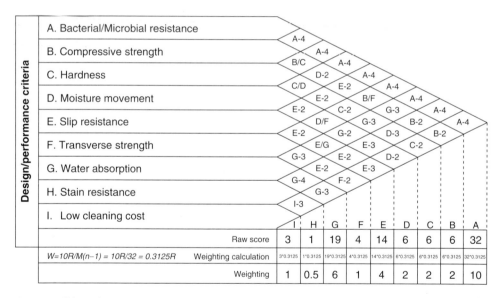

Figure 8.1(b)   Calculation of weighting from raw scores.

Whichever approach is adopted, this second stage requires each finish to be scored (on a range of 1 to 5) according to the degree to which it satisfies each of the criteria set out in the matrix. These scores are inserted in the top box of each cell of the analysis matrix, beneath the criteria matrix. Thus, for example, ceramic tiles are rated as 'excellent' (score of 5) on criteria A, G, H and I, and 'very good' (score of 4) on the remainder. Vinyl tiles are rated highly on criteria G, H and I, but as 'poor' or 'fair' on criteria A, B, C, D and E. These scores are then multiplied by the weights (or, if preferred, the raw scores) to give the entries in the lower box of each cell of the analysis matrix. For example, ceramic tiles are given a score of 24 (4 × 6) for criterion G (water absorption) and 40 (4 × 10) for criterion A (bacterial/microbial resistance).

The overall score for each finish is computed simply by adding the scores in the lower boxes across the whole row of criteria. It is then possible to proceed to the whole life appraisal exercise by selecting, for example, only the three top scoring finishes, or only those that score above a predetermined level.

In summary, what the weighted evaluation of Figure 8.1(c) does is, first, to identify the relative importance of the various criteria which the choice of finish has to satisfy in the particular functional space. Second, it shows the extent to which the various finishes satisfy these criteria, both individually and in aggregate. Most finishes are rated highly on at least one criterion, for instance, the granolithic finish has excellent transverse strength, but this is relevant only in a particular functional space.

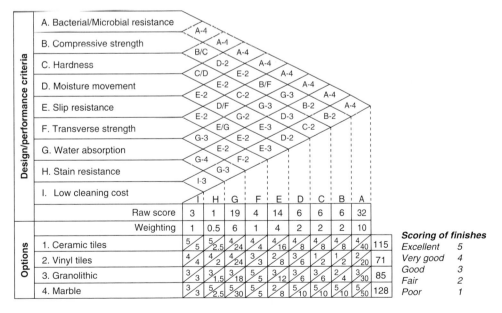

Figure 8.1(c) The completed weighted evaluation.

8.4   A whole life approach involves a feedback system

A whole life approach involves capturing historical running costs, measurement, and performance data on existing facilities/assets or systems. These data are not inexpensive to collect and analyse. As a result, information on a facility should only be collected if there is a perceived value in doing so.

What is feedback?

Feedback has been defined as the return of a portion of the output of a process or system to the input, especially when used to maintain performance or to control a system or process; this is particularly true for whole life appraisal. In the simplest form, feedback is a means of learning from experience, by carrying out the processes of reflection and deduction, i.e.

❑   analysing the experience;
❑   identifying lessons learned;
❑   generalising from these to apply the learning to other situations.

From a systems viewpoint feedback is the means by which a system compares its actual performance with an intended target. This requires four components if it is to be effective:

❑   medium – a communication channel;
❑   message – information defining or describing the state of the system;
❑   target – a means of comparing the actual state with that intended, a benchmark;
❑   action – specific action to be taken – the consequences.

For feedback to be implemented and used successfully, the methodology and motivation are of equal importance. The system has to be willing and able to carry out the feedback. Without the proper methodology and adequate motivation, feedback procedures will not actively benefit the whole process.

The problem with a systems viewpoint is that some process is very complex and it can be difficult to identify clear sub-systems. Furthermore, feedback information may not always include a specific target or performance criterion, making this view potentially restrictive. A broader definition of feedback would be relevant to design, namely: 'Feedback is the collation and dissemination of information collected at any stage of a project, which can be used to improve the design of that or future projects'.

What is the feedback cycle?

Feedback has been described as the process of learning from experience. And sometimes the process could be very straightforward such as, learning to open a bottle, or very complex, such as learning to balance an air-conditioning system. The process of learning from experiences is not readily obvious or simple. Thus, it becomes necessary to have specific feedback mechanisms in place to achieve this. The length of feedback cycle varies among different industries, from a matter of minutes, to a few hours, to less than a week, or to an even longer period of time. The feedback cycle has been developed as a generic mechanism applicable to the construction industry and it comprises four distinct steps forming a continuous loop, as shown in Figure 8.2.

Step 1   Obtain information

The first step requires the identification and collection of relevant information. Obtaining information is not a simple process,

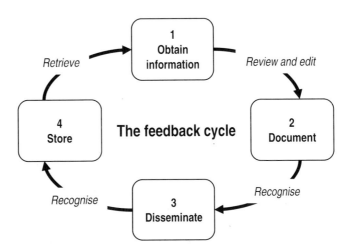

Figure 8.2 The feedback cycle.

it is the most difficult of the steps, as many barriers to the provision and collection of relevant information exist, for example, the information may be owned by others.

Step 2   Document

The second step in the process is to ensure that the information is recorded in a form that can be easily used, i.e. properly documented. This requires knowledge of how the information is to be used and some planning on the most effective way of recording this.

Step 3   Disseminate

The third step is to disseminate this information to those who would benefit from it. This could include any participants in the process.

Step 4   Storage

The final stage is the storage of the information. This may be formal, via a 'feedback file' or informal – a note. Any information that would be considered as a 'feedback' should be stored. Storage is senseless unless retrieval is straightforward. Inadequate consideration of this step will result in the absence of information for futures use. This final stage is vital for 'the feedback cycle', without it the whole cycle would be immediately disrupted and any feedback process in place rendered less valuable as a result.

Effective use of feedback has the potential to raise quality and standards in all the industries. It is an essential component of the developing process and part of continuous improvement.

8.5   Whole life analysis

Whole life analysis involves looking back to examine facilities in use. They can be of any age, any condition and may have multiple occupancy. If a 20-year old school, or a prison block is considered; the facilities would have been modified since they were first constructed, there will be new extensions that may only be a few years old, the boilers may have been replaced, and the interiors will have been revamped. In the perfect world the as-built drawings will have been updated to reflect the work undertaken; in the real world the pressure on time often means that the drawings are not up-to-date. It has become easier with CAD systems to keep information up to date.

8.6 Whole life planning

Whole life planning involves looking forward to plan the cost and performance over a time horizon. It involves analysing the proposed facility, using updated historical data, information from suppliers and specialists, and using informed knowledge to plan the cost and performance – see Figure 8.3. When considering a PFI/BOT/Concession project, the organisations owning and operating the facility over a 25-year horizon need to know the owning and operating costs, they also need to optimise the capital and running costs by comparing options; that is where discounted costs are brought to a net present value/cost. A net

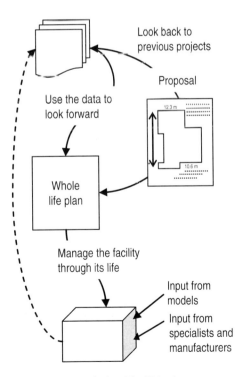

Figure 8.3   Analysis of facilities in use.

present value/cost is of no value unless it is being used for comparative purposes to make a decision. Telling a client that capital cost is €10 million and the NPV is €26 million over a 25-year time horizon at 5% real discount rate has no meaning whatsoever. Do not confuse the discounted costs with the full-year effect cost where costs are not discounted, but inflated into the future to gain a knowledge of the cash flow for the scheme.

8.7   The relationship between whole life analysis and whole life planning

Table 8.1 shows the relationship between the data requirements of whole life analysis and those of whole life planning for a facility. Historical data collected from records or drawings will be influenced by a number of variables. For example, the annual fuel cost of running a heating system can be calculated from the fuel bills. However, it is difficult to deduce the design characteristics of the facility or the occupancy pattern from these data. When using them as a price source for WLP, a reasoned approach as to whether the historical data are strictly comparable with the proposed facility is therefore required. Advice should be sought from the client on the proposed occupancy, because the forecaster has little or no control over how the facility will be used.

Exercise care when using data

A word of caution: while a facility must be seen as a complex system, there is an imperfect understanding of the full set of interrelationships and interdependencies between the components of that system. For example, in the case of window design, energy usage is not simply dependent upon the superficial window area, it is also affected by factors such as internal and external temperature differential, type of glazing, exposure, orientation, building usage, and location.

Data are available but not always in the best format

In most instances a detailed whole life analysis is likely to be undertaken only as a commission from a client. Furthermore,

Table 8.1   The relationship between the data requirements of whole life analysis and whole life planning.

Item	Whole life analysis	Whole life planning
Description of the facility, location and consultants	Required	Required
Measurement of floor, window, wall, roof areas and any detailed elemental areas as required	Measurement from drawings or on site	Measurement from drawings
Performance (annual)	Actual records (possibly more than 1 year) – use of electricity/gas – water usage if metered – security (level authorised) – cleaning frequency (windows, common areas) – maintenance (planned, repairs and annual) – grounds maintenance (details of work)	Estimated (expert advice)
Annual and non-annual maintenance – including routine maintenance, minor repairs and alteration/ adaptation/replacement	Actual records on component lives and planned maintenance (possibly more than 1 year)	Estimated
Operations cost (annual)	Actual records (possibly more than 1 year)	Estimated
Condition of facility	Site visit to determine condition of elements. Use any condition surveys that are available	As specified
Discount rate, inflation rate, whole life	Not applicable	Provided by the client/ owner

consultants will not have access to operations and maintenance cost records unless they are engaged on a project for the client: the problem of access to the data is a key issue. For example, in the further and higher education sectors, detailed records are kept on maintenance of their facilities for the purpose of costing their expenditure and for budgeting in subsequent years. But, the data are in a format to suit the costing system used by the college or university; not in a format as feedback on performance in use.

8.8   Cost relationships

Some mention has already been made of the influence of the way a facility is used and its impact upon the running costs. Data on facilities used for 24 hours a day cannot be compared with 10–11-hour occupancy cycles, or those with high staffing levels, such as security and caretaking staff, and must be recognised as such.

The tendency when relating any aspect of cost, is to use a unit of measurement as the basis for comparison, either as a cost per m^2 of the gross floor area (gfa) or as a cost for the relevant unit quantity. This practice is also followed in whole life cost. The whole life cost components can be considered as:

❏   Area related – where the cost is related to the area, for example, cleaning.
❏   Use or function related – where the user is the predominant influence on cost, for example, porterage and caretaking.
❏   Non-area related – for example, the cost of a plant engineer to maintain the plant and equipment.
❏   Price related – where the cost is related to the price or replacement value, for example insurance, local taxation.

Clearly these are also affected by the physical characteristics. A simple example illustrates this point. The annual running costs of a heating system for a $5000\,m^2$ facility may be assumed to be as follows.

			Cost
1	Fuel	Area-related and part user-related	150 000
2	Maintenance contracts on boiler and pumps	Non area-related	10 000
3	Insurance on services	Non area-related	5 000
4	Replacement (provision)/sinking fund	Non area-related	15 000
5	Plant engineer	Non area-related	40 000
			€220 000
			$= €44/m^2\,gfa$

A large proportion of these costs is not area related, limiting the usefulness of the measured cost per unit area. If a plant engineer had not been considered necessary the cost would have been €36/m² gfa, the assumption being that the caretaker would be responsible for start up and monitoring. Thus, in the absence of detailed information there is the possibility of data being erroneous if based solely on a unit price rate.

It is easiest to think of the costs in terms of fixed and variable costs. The fuel cost will be a variable cost related to the floor area and the use characteristics, whereas the plant engineer is a fixed cost. Many of the fixed cost items will have some relationship to the initial capital cost of an item. Insurance premiums are generally related to the replacement costs plus an allowance for professional design team fees.

Maintenance work can be divided into annual and intermittent work. Care should be taken when using historical cost data for this work. For example, redecoration of facilities is usually an ongoing process. In practice the planned redecoration cycles are often not followed owing to differences in climatic exposure, amount of use, and opinion as to acceptable standard. Facilities are often decorated piecemeal to avoid undue disturbance to the occupants. Therefore, when a redecoration cycle is not followed rigidly, the historical cost records for the past three years may not capture the cost of complete internal and external redecoration.

Figure 8.4 illustrates the assumptions that might be made for the redecorations cost category in a whole life plan. There are two threads to be followed:

1   Use current costs with an allowance for inflation in the future for the purposes of budgeting and planning.
2   Use discounting to a net present cost/value. Discounting to today's cost is only useful if options are being compared, say the choice between emulsion paint or fabric and wallpaper on the walls.

The approach is as follows:

❑   Estimate the superficial area and initial capital cost for the decorations.
❑   Estimate a cost for the annual periodic inspection and making good of the decorations.
❑   Determine a redecoration cycle for the various parts of the facility.

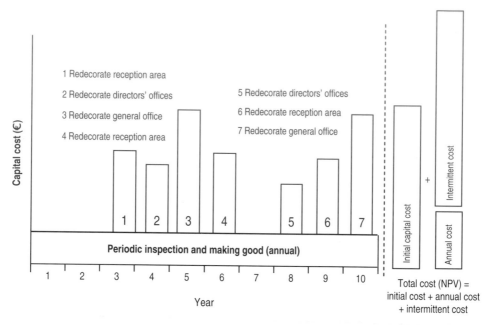

Figure 8.4 Redecoration costs.

❏ Calculate the superficial areas and estimate the cost of redecorating on the above cycle at current costs of labour and material.

If options are being considered:

❏ Discount future costs to a net present value (where all cash amounts are converted to an equivalent value occurring now) or to an annual equivalent value (where all cash amounts are converted to a time equivalent value occurring in a uniform amount each year over the whole life).

8.9 The sequence of whole life analysis, whole life planning and whole life management

While whole life analysis, whole life planning and whole life management can be viewed as separate activities, a logical sequence which links them together. Table 8.2 shows this sequence, where the assumption is that data for a whole life plan of a proposed facility are based upon three similar buildings for which analyses are available.

Table 8.2 The sequence linking WLA, WLPA and WLM.

Step	Actions
1	Get quantitative and qualitative data ❑ Measure the areas ❑ Decide upon the performance requirements and the categories to be used for the whole life plan
2	Gather data from past projects and from specialist organisations ❑ Select whole life analyses from past projects similar in size and type to the planned facility ❑ Consider how to aggregate the information, use benchmarks to ensure reliability ❑ Consider how to update the information to current prices using indices ❑ Decide what reliability can be placed on the data ❑ Cross-check figures with specialist contractors and with models
3	State assumptions ❑ State all assumptions being made. State exactly what has been assumed and why
4	Using a spreadsheet ❑ Put all the information into a spreadsheet, such as areas, unit rates, inflation rates, engineer's estimates for energy use, maintenance cycles, degradation of materials, energy consumption, cleaning cycles and insurance
5	Select the time horizon for the analysis ❑ The time horizon is dependent upon the client's perspective. For a PFI project, it may be 35 or 30 years; for a developer it may be 10 years; for a homeowner it may be 15 years. It is usually the service life of the facility. For example, if a bridge has a service life of 70 years, it would be unrealistic to try and forecast over its 70-year life
6	Extrapolate into the future ❑ For a whole life plan, decide upon the service/design life of components, obsolescence and renewal ❑ Decide what annual, cyclical and periodic maintenance will be required ❑ Decide how inflation should be handled, either as one rate in the discount rate, or whether different elements will inflate at different rates ❑ Produce a whole life plan
7	Discounting ❑ If different options are being considered, discount future costs over the time horizon back to a net present value/cost. This enables the lowest cost option/best value to be identified
8	Residual values ❑ Consider what value the facility or component may have as scrap or reuse. Often the residual value is in the land the facility uses
9	Plot the results ❑ Check the sensitivity of decisions by testing with different rates
10	Check the sensitivity ❑ Monitor and modify the whole life plan through the design and construction phase into occupancy

(Continued)

Step	Actions
11	Monitor and modify ❏ Monitor and modify the whole life plan through the design and construction phase into occupancy
12	Set up a framework and a system to collect operational and cost information for whole life management ❏ The data can be used for budgeting and for making informal decisions about the maintenance plan, energy costs, and overheads
13	Feed back information on performance and cost to the design team

Indices

The principles for using index numbers will be familiar, but it should be noted that there are several problems in applying index numbers to cost data. First, an index number is an average. The weighting used in constructing the index number may not be precisely appropriate to the project in hand. There is no easy way of overcoming this problem. Users should exercise discretion and be as fully informed as possible about the underlying construction of the index they are proposing to use. Second, different cost factors tend to inflate at different rates. Thus it may be preferable to use a number of individual cost indices, for example, for labour, energy, and cleaning, rather than an aggregate index of costs.

8.10   Documentation format for whole life planning

Figure 8.5 shows a format for a whole life plan. The data are a combination of the following:

❏ General/descriptive
❏ Performance
❏ Measurement
❏ Assumptions
❏ Cost

These data should be used in such a way that the design team can evaluate total costs or the total cost consequences of design options. Figure 8.5 is only one of many possible formats for the data, and everyone will have their own views on how best to display the information

SECTION 1 General information

Project...

Location...

....

Date plan prepared..

Date for price base..

Discount rate...

Time horizon...

Person producing the plan.............................   ....

Sources of data used in the plan	Description of proposed facility	Assumptions (list all assumptions –
List sources where appropriate	**or system**	be as comprehensive as possible) Occupancy Obsolescence Design/service life

SECTION 2 Facility data (*use separate sheet for any elemental breakdown required*)

Gross floor area ....m^2	Windows and external door area	Fittings and furnishings ....No
Circulation space area ....m^2	(net) ....m^2	Sanitary appliances ....Nr
Ground floor area ....m^2	Internal walls and partition area ....m^2	Water installation (draw off points) ....Nr
Volume (if applicable) ....m^3	Internal door area ....m^2	Site works ....m^2
Upper floor area ....m^2		
Roof area ....m^2	Wall finishes area ....m^2	
Wall area (net) ....m^2	Ceiling finishes area ....m^2	

Design data	*Give brief specification*
No. of occupants	
Occupants/m^2 gross floor area	
Net/gross floor area	

Figure 8.5   Whole life plan.

SECTION 3 Initial capital costs

	Total	Per m² of floor area		Total	Per m² of floor area
Capital cost			Furniture and equipment		
Building			Land (if applicable)		
External works					

SECTION 4 Annual costs

	Total to time horizon	NPV factor	NPV		Total to time horizon	NPV factor	NPV
Cleaning – internal – external – windows				Operations			
Energy – heating – hot water – lighting – power – other				Insurances – structure – contents – employer's liability – consequential losses			
Local taxation				Gardening			
Utilities – water – sewerage – telecoms				Salvage and residual value – demolition – re-sale/scrap value – other			
Facilities management				Security			

SECTION 5 Intermittent costs

Redecoration				Replacement			

(Continued)

Section 1 (of Figure 8.5) gives general and descriptive data about the proposed facility together with assumptions. Section 2 lists facility measurement data, all of which are likely to have been used in the preparation of a price forecast or capital cost plan for the proposed facility. Areas should be based upon the measurement rules used in the elemental categories. Mention is made of an additional section for any elemental breakdown. For example, part of the ceiling finishes area might be suspended ceilings and part painted fair face concrete. In such situations, a breakdown of the areas would be useful because of different operations and maintenance requirements.

Section 4 details annual operations and annual maintenance costs and Section 5 gives the intermittent costs. Account should be taken of area-related and non-area-related costs when building up prices. Information is needed on the intermittent maintenance costs, which consist of the replacement cost of each item at current levels of labour and materials together with the replacement cycle. In most situations, only a proportion of an item will be replaced completely at a specific time.

Where appropriate, all the running costs calculations should be adjusted to a net of tax figure, bearing in mind that the running costs of a facility are allowable as a business expense. This will be particularly relevant for trading organisations.

8.11   Costs and benefits

Whole life appraisal should also take account of any non-monetary costs and benefits. Non-monetary costs are, for example, delays caused by traffic arising from roadworks.

Table 8.3 is the summary section for the whole life plan.

The present value column of Table 8.3 is used where alternative options are being considered. If the purpose is to plan through a time horizon, then adjustment should be made for inflation using full year effect costs.

Table 8.3   Whole life appraisal summary.

Project ....

Location ....

Date ....

Costs	Target costs	Target cost/m² gross floor area	Present value
1 **CAPITAL COSTS** e.g. Substructure Superstructure Finishings Fittings and furnishings			
TOTAL CAPITAL COSTS			
2 **RUNNING COSTS** 2.1 Operations costs e.g. Energy Cleaning Local taxation Insurances Security and health Staff Management and administration Land charges			
TOTAL OPERATIONS COSTS			
2.2 Maintenance costs (annual) e.g. Main structure Internal decorations External decorations			
TOTAL MAINTENANCE COSTS			
2.3 Maintenance/Replacement/ Alteration Costs (intermittent) e.g. Main structure Internal decorations External decorations			
TOTAL MAINTENANCE/ REPLACEMENT/ALTERATION COSTS (INTERMITTENT)			
2.4 Sundries			
Total sundries			
Total running costs			
3 **ADDITIONAL TAX ALLOWANCES**			
Total add. tax allowances			
4 **SALVAGE AND RESIDUALS**			
Total salvage and residuals			
Whole life cost			
Total present value of whole life costs			

9 Taxation and whole life appraisal

9.1 The implications of taxation and grants for whole life cost

The taxation aspects of construction can have a fundamental effect on the true costs of the construction and operation of facilities. Effective tax planning can turn an apparently non-viable project into a viable one, or change the ranking order of a number of project proposals or design choices. Even in the absence of such dramatic effects, effective taxation cost planning offers significant benefits for clients of the construction industry. This is so because the greater the proportion of the cost eligible for tax relief, the lower the net (after tax) cost will be.

> Most PFI projects are now being structured to allow capital allowances to be claimed by the special purpose vehicle providing the asset. Many of the projects are potentially very rich in allowances and with careful planning as much as 40% of the cost can be claimed in some instances.

This chapter examines a number of the more important elements in the taxation of facilities in the UK. However, tax regulations change from year to year, not merely in respect of rates of tax that are applied, but also in terms of the types of facility that will be eligible for particular tax allowances. Those who seek to tax plan facilities must become familiar with the relevant Finance Acts and other regulations, and the distinction between capital expenditure and revenue expenditure must constantly be borne in mind.

Capital expenditure is a term used to refer to money expended in acquiring assets, or in the permanent improvement of, or the addition to, or the extension of, existing assets, which are intended for use in the carrying out of business operations. These assets should be expected to have a useful life of more than one year.

Revenue expenditure is expenditure charged against expense in the period of acquisition. This term is used to refer to all expenditure that cannot be debited to an asset account.

In the calculation of a company's liability to tax, revenue expenditure is deductible before settling the amount of profit liable to tax. In the context of whole life appraisal, the items that are considered a business expense are rent, energy, cleaning, local taxation, water, lighting, security, insurances, staff, maintenance, and management expenses of the facility, repairs to the facility and plant, costs of renewal of plant, and writing down of obsolescent plant. In general, it may be said that any expenditure incurred as a running cost of a facility is deductible against liability for tax.

Capital expenditure, on the other hand, is not treated as a simple business expense for tax purposes. Expenditure on new facilities or on the enlargement or improvement of an existing facility is considered as capital expenditure and cannot be deducted directly in calculating tax liability. The principle appears to be that if capital receipts are not subject to income tax, then capital expenses are not allowable. In some instances the cost of renewing worn out capital items of plant and machinery is allowable as a running cost at the time of renewal, thus providing the one exception to the principle.

However, certain types of capital expenditure are eligible for capital allowances. The nature of these capital allowances and their whole life cost effects are analysed below.

9.2   Capital allowances for facilities

Capital allowances are available only on selected types of facility, but then the majority of the allowances can be taken in the early life of the facility. Thus, it is to be expected that facilities will be written down for tax purposes much more quickly than for accounting purposes. The main reason for this is that while a company in reporting to its shareholders is interested in calculating its trading profit and balance sheet as accurately as possible, the government is more interested in regulating investment. The underlying philosophy is that the government wants companies to invest in plant and equipment to improve the nation's economic performance.

Where a facility qualifies for capital allowances, the general structure of these is as follows:

- ❏ Initial allowance.
- ❏ Writing down allowance.
- ❏ Balancing allowance.

The initial allowance is made to a company in respect of the construction cost. For plant and equipment the writing down allowance is on a 24% reducing balance, dependent upon the tax rules in force.

The writing down allowance uses the principle of year-on-year write down. Take as an example an industrial building with 4% straight line writing down allowance, so that the allowance was spread evenly at 4% per annum over 25 years. This meant that at the end of the 25 years, the full capital expenditure would have been allowed against profits.

The balancing allowance is made if the facility is sold for a sum exceeding the written down allowance. However, if it is sold for a lower figure than its written down allowance, then a further allowance is given to make up the difference.

The reducing balance allowance works as follows. If the plant and equipment were being written down on a 25% reducing balance; in the first year the value would be reduced by 25%. In the second year it would be 25% of the remaining 75% (i.e. 18.75% of the original cost), in the third year it would be 25% of the remaining 56.25% of the original cost (i.e. 14.06% of the original cost) and so on. Capital allowances for assets which qualify as plant and equipment are assessed using the 'pool concept'.

There need be no direct connection between depreciation allowances entered in the profit and loss account or the balance sheet, and the depreciation or capital allowances taken for taxation purposes. In an accounting context, depreciating an asset means spreading the cost over its estimated useful economic life. For example, this might be by the straight line or linear method of depreciation, which involves dividing the cost, less estimated scrap value, of an asset by the estimated economic life. The charging of depreciation simultaneously reduces the recorded amount of the fixed asset concerned in the balance sheet and the net profit of the company.

Essentially in the UK, capital allowances are available on new industrial buildings, hotels, agricultural buildings, and small workshop buildings. To determine whether an industrial building qualifies for capital allowances, and the precise form of such allowances, consideration must be given to the type of trade and the nature of the building.

Industrial and agricultural buildings attract a writing down allowance of 4% of cost per year. If part of the building is used for domestic or private purposes, the allowance stands on the whole cost if the non-industrial part is 25% or less of the cost of the whole building. Land owned by the business does not qualify.

A sports pavilion used by a trader for the welfare of his employees is treated as an industrial building whether or not the trader is carrying on a qualifying activity. In addition, a building may be treated as an industrial building for some years and not for others, if it is used for different activities at different times.

The word structure embraces artificial works which might not properly be described as buildings, such as wells, bridges, roads, culverts, tunnels, and so on. Expenditure on the land itself is excluded from the allowance. However, the cost of demolition and the professional fees of the design team are eligible for capital allowances. Furthermore, the extension to an existing industrial building will qualify in the same manner as a new industrial building.

In 2004 the corporation tax main rate is 30%. The small companies' rate is 19% for companies with taxable profits between £50 000 and £300 000, and the starting rate is zero for companies with taxable profits of £10 000 or below. Marginal relief eases the transition from the starting rate to the small companies' rate for companies with profits between £10 000 and £50 000. The profits limits may be reduced for a company which is part of a group or has associated companies. The lower rates and marginal reliefs do not apply to close investment holding companies.

> The whole life cost effects of these capital allowances are significant as can be seen, for example, on a new industrial building, excluding land that costs £1 000 000. In the year of purchase, capital allowances amount to £790 000 (the initial allowance of 75% and the first year writing down allowance of 4%). With Corporation Tax at 30%, the client will reduce his overall tax liability in the purchase year by £237 000, while the net present value of the total tax savings (assuming a net of inflation discount rate of 5%), will be approximately £280 000.

The impact on design choice of the capital allowances can also be significant. For example, in the case of a company involved in the storage, distribution and retailing of goods such as food, if the storage areas are located within the retailing units no capital allowances will be given. On the other hand, if the storage areas are combined as a distribution centre within a qualifying industrial unit, capital allowances are permitted.

9.3   Capital allowances for machinery, plant and equipment

Expenditure on qualifying plant and machinery is normally given at a rate of 25% a year on a reducing balance basis and is deducted from taxable profit. This allows the benefit to be spread over a number of years. For example, for a 30% corporation taxpayer, £100 expenditure incurred on qualifying plant and machinery would equate to a £23 tax saving over the first five years.

Allowances are given for capital expenditure on machinery or plant used for the purpose of a trade, profession or employment. The expression 'machinery and plant' is widely interpreted for tax purposes, it includes machinery in the general sense and also such items as typewriters, desks, office equipment, carpets, curtains, demountable partitions, shop counters, and electrical fittings, fixed or moveable, which are kept for permanent employment in the business.

Items in a facility that will frequently qualify as plant and equipment are lifts and escalators, air conditioning, burglar alarms and security equipment, boilers and hot water equipment.

Expenditure on repairs and maintenance or on renewal of parts, does not qualify for any of the capital allowances, but is considered to be a revenue trade expense and so is deductible.

Capital expenditure on alterations to an existing building, incidental to the installation of plant or machinery, may be treated as though it were expenditure on that machinery and so may qualify for the allowances. For example, if a new item of plant required a concrete machine base, drainage, and electrical connections, all the ancillary work including the excavation and concrete would qualify as plant because it would not have been necessary but for the new plant.

9.4   Capital savings allowances and energy savings

Businesses will be able to claim 100% first-year allowances on their investments in designated energy-saving plant and machinery. Investments can qualify for enhanced capital allowances following the publication of the Energy Technology List. Under the scheme, businesses that invest in qualifying energy-saving technologies will be able to write off immediately the whole cost against their taxable profits of the period during which they make the investment.

Qualifying technologies

The scheme will apply initially to the following designated technology classes, provided the equipment meets strict energy-saving criteria:

- Combined heat and power (CHP)
- Refrigeration equipment
- Boilers and add-ons
- Thermal screens
- Motors
- Lighting
- Variable speed drives (VSD)
- Pipe insulation.

Examples of the tax implications on a PFI project

(Source: Inland Revenue)

Example 1

A private sector 'operator', whose trade includes the provision of design, construction and maintenance services, enters into a Private Finance Initiative (PFI) contract with a public sector 'purchaser' to build student accommodation and maintain it for 30 years. The trade commences when the PFI contract is signed. In return the operator receives an annual service payment, the 'unitary charge', which commences after the accommodation is completed.

Accounting period 1

The accommodation is completed at the end of the first accounting period. For tax purposes the design and construction costs are revenue expenditure. The accommodation is not a fixed capital asset of the operator's trade. For accounting purposes the example assumes that it is reported as a fixed asset on the operator's balance sheet at a figure of £90 m representing cost.

	Debit		Credit
Fixed asset	£90 m	Bank	£90 m

No income is received in the first accounting period.

Accounting period 2

In the second accounting period a unitary payment of £15 m is receivable. For tax purposes the £15 m is trading income for the

	Debit		Credit
Bank	£15 m	P&L account	£15 m
P&L account (depreciation)	£3 m	Accumulated depreciation account	£3 m

Note: For tax purposes we follow the accounting recognition of income and expenditure in the profit and loss account, subject to any relevant over-riding statutory or case law principle.

provision of services. For accounting purposes the whole of the unitary payment is credited to the profit and loss account. Depreciation on the fixed asset, calculated at £3 m, is debited to the profit and loss account.

The £15 m unitary payment is trading income for services provided. The £3 m depreciation represents revenue construction expenditure and is an allowable deduction for tax purposes. Therefore no adjustments are required in the Schedule D Case I computation.

Computation

Income (net of depreciation) £12 m
Profit (before overheads) £12 m

Example 2

A private sector 'operator', whose trade is running a prison, enters into a Private Finance Initiative contract with a public sector 'purchaser', to provide a specific number of prison places for 30 years. The operator builds a prison on land acquired for the purpose. The trade commences when the prison is completed and ready to receive its first prisoner. In return the operator receives an annual service payment, the 'unitary charge', which commences after the trade has started.

Accounting period 1

The prison is completed at the end of the first accounting period. For tax purposes the design and construction costs are capital expenditure. The prison is a fixed capital asset of the operator's trade. For accounting purposes the example assumes that it is reported as a fixed asset on the operator's balance sheet at a figure of £90 m representing cost.

	Debit		Credit
Fixed asset	£90 m	Bank	£90 m

No income is received in the first accounting period.

Accounting period 2

In the second accounting period the trade commences and a unitary payment of £15 m is receivable. For tax purposes the £15 m is trading income for the provision of services. For accounting purposes the whole of the unitary payment is credited to the profit and loss account. Depreciation on the fixed asset, calculated at £3 m, is debited to the profit and loss account.

	Debit		Credit
Bank	£15 m	P&L account	£15 m
P&L account (depreciation)	£3 m	Accumulated depreciation account	£3 m

For tax purposes we follow the accounting recognition of income and expenditure in the profit and loss account, subject to any relevant over riding statutory or case law principle. The £15 m unitary payment is trading income for services provided and no adjustment is required in the Schedule D Case I computation. The £3 m depreciation represents capital construction expenditure, is not an allowable deduction for tax purposes, and is therefore added back in the Schedule D Case I computation. Capital allowances can be claimed on qualifying expenditure.

Case I computation

Income (net of depreciation)	£12 m
Plus depreciation	£3 m
Profit (before overheads)	£15 m

10 Integrated logistic support (ILS)

10.1 What is ILS?

Integrated Logistic Support (ILS) is a management tool traditionally used in the defence sector to define the support requirements of equipment. It helps to procure a lean operation, aiming to supply only as much support as is needed, at the time it is needed. It ensures that the design of a facility and its components includes consideration, over its whole life, of its performance, maintenance, operating characteristics and costs as well as support when it fails.

Design for support . . . then support the design

10.2 Where did ILS come from?

In the mid-1970s the US Department of Defense supported a private venture to establish a strategy where the interactive effects of support decisions could be analysed. This allowed 'what if' calculations to be carried out, which, when linked to cost models, provided a competitive assessment between support alternatives. The concept evolved and became a national military standard, initially operated through a 'card system' database. Without ILS, there would be no back-up for armaments, no fuel or spare parts for aircraft on operations, and no supply of food or clothing for the troops (Boyce, 2002).

The UK Ministry of Defence (MoD) has introduced ILS into defence equipment procurement and developed a standard that prescribes requirements for the application of ILS. Defence

Standard 00–60 provides the disciplines for ensuring that supportability and cost factors of equipment are identified and considered from concept, and throughout their lifetime. The overall aim is to provide defence equipment at optimum whole life costs.

The use of ILS has widened into other industries, where the importance of whole life appraisal has been recognised, for example the aircraft and petrochemical industries. Some industrial areas that operate in synergy with MoD procurement principles have also recognised the financial benefits that can be attained through the introduction of ILS. For example, the UK Civil Aviation Authority has fully endorsed the use of ILS for their major procurement projects. Other industrial sectors, such as vehicle users (from buses to Formula 1), have also introduced ILS into their supply chains.

10.3   What is the difference between ILS and whole life appraisal?

Whole life appraisal focuses on cost and performance of a facility or element over its life. It aims to balance cost and performance through design, construction, and operation. ILS takes this one step further and includes the support for that facility or element over its life. It considers how and why components and equipment fail and their likelihood of failure. It looks at how and when the repairs should be undertaken. The basic principle of ILS is that all capital goods must be maintained and operated over their service life; there are two issues (1) what needs to be done and when, (2) how the work should be done and by whom, which may include training the workforce to maintain the equipment, and consideration of the logistics of obtaining parts and equipment.

10.4   ILS and construction

A management and technical approach which ensures that the owner/user will receive a facility that will meet performance requirements (durability, reliability, maintainability, performance, availability) with a lowest whole life cost.

Support means maintenance, repair, refurbishment, and the way in which a facility or component is kept at optimum performance – including breakdown and failure. It is a systematic management process with an important role in ensuring that the organisation's objectives are met while presenting the optimum financial commitment for the facility. The process provides a suite of techniques for planning, scheduling, control and integration to achieve the optimum operation and maintenance goals to support the functional requirements of occupiers.

A UK strategic forum for construction reported that there is a considerable amount of waste incurred in the industry as a result of poor logistics. BAA, the UK's major airport operator and owner is one of the firms pioneering the use of logistics on construction projects. They opened a supermarket style 'consolidation centre' to receive materials designated for works on the airport. Materials are distributed to sites at night ready for use the following day. Further, on the Canary Wharf development, where there are 300 vehicles delivering to twelve high-rise projects each day, a logistics team is using 'zone management' software to synchronise deliveries to suppliers using pre-allocated slots on 70 hoists and tower cranes.

10.5 Why use ILS?

The procurement process in construction remains traditionally based on a capital cost approach, despite the theoretical understanding of whole life appraisal for over twenty years. It is only recently that WLA has started playing a significant part in decision making. The demand for excellence is changing the criteria; government policy now requires that public procurement is based on value-for-money – that being 'the optimum combination of whole life costs and quality to meet the users' requirements' (Sustainability Action Group of the Government Construction Clients' Panel, 2000).

The growth in the Private Finance Initiative (PFI) and Build Operate and Transfer (BOT) procurement approaches necessitates an awareness of whole life appraisal, prompted by owners becoming concerned at the high occupancy and maintenance costs of their facilities – the cost of ownership. The organisations with a 25-year concession to own, operate, and maintain a road want to ensure the operating and maintenance costs have been properly planned, budgeted, and implemented. Assuming an annual UK construction output of €120 billion, facilities have to be maintained, repaired, heated, cooled, cleaned, insured, and managed throughout their service life. Assuming that 25–40% is in the capital spend, a conservative estimate is that the whole life cost of a €120 billion investment is closer to a €400 billion spend over the whole life of the facilities. Some facilities, such as tunnels and bridges will be operational into the next century!

10.6   The concepts of ILS

ILS concepts do not imply a complete revolution, discarding all existing systems and procedures and starting afresh; rather, ILS is a catalogue of techniques, principles and standards to enable adoption of good practice and the sharing and exchange of information. This sharing makes the process more transparent and allows the cost of ownership to be visible. Most costs are fixed during the design process and cannot be altered easily afterwards. A large proportion of inputs during the construction process is repeated from facility to facility. Much of the repair and maintenance is also repetitive. Aspects of facilities such as mechanical and electrical engineering services and other subsystems are mass-produced. ILS provides an integrated suite of techniques to mould into whole life appraisal for decision making at the design stage of a facility and throughout its whole life. ILS is applied in more detail as design develops, is refined, and moves through to detail design; this both influences the elemental design and sustains an ongoing whole life appraisal model.

10.7   The main steps of ILS

The four main steps of ILS are:

1   Ensure that whole life support considerations influence the selection and design of components. Reality has to prevail as at the early design stages the design will still be evolving, ILS will become more detailed as more information becomes known.
2   Develop integrated support requirements compatible with the objectives of the components, for example, in the case of a lift installation any repairs will be critical as the lifts may become dangerous or render the facility unusable, whereas a failure of the cooling system may mean discomfort, but not critical failure of the facility.

3   Acquire the necessary support.
4   Provide support for optimising whole life cost.

10.8   The benefits of ILS

ILS benefits owners, users and organisations across the construction supply chain. The owner should ensure that the design and construction team is aware of the high priority placed on minimising the running costs and managing them, rather than simply monitoring them.

The cost of the ILS process should be charged to the project. The cost of providing the service is likely to be minimal once the system is established. The ILS tasks to be undertaken by the contractor during design, development, construction and throughout operational life must have adequate contractual cover. To reinforce the logistic aims in the owner's brief, the contractual provisions must clearly state that logistic matters have equal priority with operation/performance. The contract must be framed such that it makes the price of failure in the logistic support area significantly high.

Resistance to change is inevitable; 'we've always done it like that' is a strong argument to avoid change; ILS requires a change of culture. There is a need to determine the core skills and identify the new skills to apply ILS. The added value of identifying the cost of ownership should become a competitive advantage.

10.9   An example of ILS

Consider a lift system where a traditional maintenance plan has been devised either at the design stage or at the occupancy/operation stage. The plan will set out repair and maintenance schedules and identify the health and safety requirements according to the Construction (Design and Management) Regulations 1994 – CONDAM. However, it is vital to establish a systematic approach in order to identify the projected costs involved, rather than just the immediate cost of the repair. The impact and the cost of the fault on the rest of the facility also has to be calculated.

Figure 10.1 shows the reliability prediction for the main motor in a lift hoist room. The report is an output of MechStress,

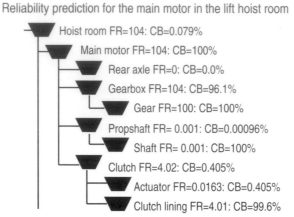

FR = Failure rate
CB = % contribution

Figure 10.1   Report from the MechStress software

software that calculates reliability prediction and the failure analysis of a system – in this case a lift. The program calculates not only the failure rate of a component but also its contribution to the whole system. For example, the clutch lining is a major component of the clutch with a contribution of 99.6%, yet its failure rate is only 4.01. This figure suggests that while the lining is a vital component, the likelihood of its failure is very small, compared with, say, the gearbox and understanding the likelihood of failure and its impact is important to the operator of a building. At the design stage this type of information can (and should) affect the decision-making process on types/quality of components.

10.10   Summary for ILS

This book is about understanding whole life appraisal as a technique and a tool. ILS has been introduced to the reader to describe how WLA optimises cost and performance, whereas ILS is about how the facility should be maintained and by whom. It embodies failure information and details the likelihood of failure. Both WLA and ILS techniques are interrelated; it is beyond the scope of this book to describe all the attributes and details of ILS. Be aware of its existence and be aware that ILS will grow in importance through facilities management operators and the logistics operators.

11 Summary and conclusions

Whole life appraisal is not a passing fad; the techniques are tried and tested. The challenge is the collection of reliable data about cost and performance of facilities/assets in use that can be fed back and used by the design team in the appraisal of new and existing facilities. The book does not contain a large number of examples of the mechanics of undertaking a whole life appraisal; there are many examples in books and reports describing the techniques of discounting and undertaking whole life cost calculations. A whole life approach is essential for making informed decisions about long-term durable assets. In looking at the cost of a facility/asset, it is short-sighted to consider merely the initial acquisition costs. Attention must also be paid to the subsequent running costs associated with the operation and maintenance of components. It is no longer appropriate to argue that because initial capital costs are the most important costs of purchasing and operating a facility, good decision making need only consider capital costs. The previous chapters have shown that a trade-off exists between capital costs, the performance of the facility in use, and running costs. There are important cost savings to be made by shaking loose from the traditional concern purely with capital costs.

A whole life approach is not solely applicable to the efficient design of new facilities. The approach offers important cost savings when applied to decisions with respect to existing facilities/assets and when applied to the choice of individual components. Any area of design, operation or maintenance that has associated costs over time will benefit from a whole life approach.

As a decision-making tool then, a whole life approach offers significant benefits to clients of the industry, but this is not its only use. Whole life appraisal should be seen as a management tool. It can be used to identify short-term running costs of facilities or components for cash flow purposes. It should also be used to monitor running costs, since by doing so it will be possible to identify areas in which savings might be achieved either by a change in operating conditions or by changing the component.

The various uses of whole life appraisal and the basic methodology of the approach are summarised in Figure 11.1. These comprise four main components – data, a series of techniques, a system for applying techniques to data, and output. The output takes a number of forms depending upon the objectives – whole life analysis, planning or management.

Whole life appraisal uses tried and tested economic principles long applied to investment appraisal in many spheres of industrial and commercial activity. A number of good reasons can be advanced as to why organisations have been slow to implement whole life appraisal:

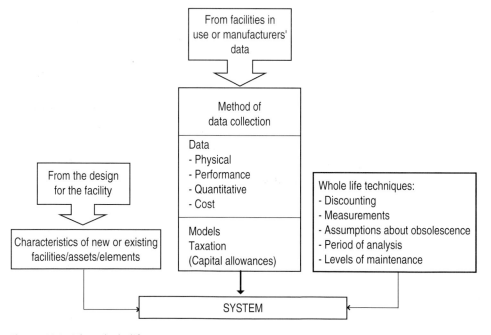

Figure 11.1   The whole life system

- ❑ The diverse nature of clients and their motivations/expectations.
- ❑ The complex and theoretical relationship between money now and money spent or received in the future.
- ❑ Changing rates of inflation over time.
- ❑ The long time lag between the design process and data becoming available on the running costs and the performance characteristics of facilities or systems.
- ❑ A lack of reliable and adequate data on running costs and performance, structured in a useful format.
- ❑ Difficulty in forecasting future events over the life of a facility and its components.
- ❑ The sensitivity of future estimates to changes in taxation and legislation.
- ❑ The natural concentration by organisations on services for which they are paid and, therefore, those in demand.

This book has addressed the issues stated above and supplies a system with the following attributes:

- ❑ It focuses attention upon the relationship between the capital and running costs.
- ❑ It provides a methodology and framework to enable the design team to estimate the total cost of what could happen, what should happen and what will happen.
- ❑ It provides a checklist of occupancy cost items to encourage the design team to bridge the gap between the design and construction phase and the operational phase.
- ❑ It uses information for a whole life cost management system for new or existing facilities that will identify those areas in which the costs might be reduced, either by a change in operating practice or by replacing the relevant system. It is important to the design team that there is feedback on the cost and performance of the facility in use.
- ❑ It structures the data, allowing a more coordinated approach to capturing cost information and performance data for both new and existing facilities.
- ❑ It evaluates the total costs of design options in a simple manner.

No mention has been made of an appropriate scale of charges for providing whole life advice, nor has the question of professional liability been discussed. It is beyond the scope of this book to discuss the legal and insurance problems of professional liability.

It should be emphasised, once again, that while the results of the analysis may look very precise and convincing, there is no substitute for professional skill and judgement when applying whole life techniques. The techniques are purely a tool for decision making.

Glossary of terms used in whole life appraisal

Annual equivalent
The present value of a series of discounted cash streams expressed as a constant annual amount.

Appreciation rate
Nominal rate of which assets increase in value over time.

Baseline date
The starting point for the whole life cost analysis, beyond which decisions deal with future courses of action. It is 'today' in the analysis. May be referred to as the baseline year (or analysis year).

Bill of quantities
A document that details the items of work involved in the construction of the project for the purposes of tendering and contract administration.

Break-even analysis
A procedure for evaluating alternatives in terms of a common unknown variable that will make the cost equations for the alternatives equivalent; this value is the break-even point.

Capital cost
The initial acquisition cost of the land and building.

Construction cost
The cost of labour, material and plant involved in the creation of the building, and other improvements to the land, including all supervision, profit and rise and fall during the construction period.

Cost–benefit analysis

The technique for the comparison of tangible and intangible costs and benefits over time for alternative projects based on either a social or economic approach.

Cost control

The overall process of budgeting, optimising, documenting, monitoring and managing the whole life costs of a building project, commencing with the decision to build and concluding when the cost implications of the project are no longer of concern to the owner.

Degredation

Changes over time in the composition, microstructure and properties of a component and material which reduces its performance.

Depreciation

An accounting device that distributes the monetary value (less salvage value) of a tangible asset over the estimated years of productive or useful life. It is a process of allocation, not valuation.

Design life

Service life intended by the designer, which is often the expected service life.

Discount rate

A rate used to relate present and future money. This is expressed as a percentage used to reduce the value of the future (tomorrow) money in relation to present (today) money, to account for the time value of money. It reflects the fact that money spent or received in the future is worth less than money spent or received in the present, since there is no interest income on that future money. The discount rate may be the interest rate or the desired rate of return. If it considers the impact of inflation of future costs, it may be called a *real* or *combined discount rate*.

Discounted cash flow analysis

The technique for assessing the return on capital employed in an investment project over its *economic life* or suitable *time horizon*, with a view to prioritising alternative courses of action that exceed established profitability thresholds.

Discounted present value
The opportunity-adjusted *present value* – this ignores changes in base worth and is thus an over-simplification of the *time value of money*.

Discounting
The theoretical adjustment for the *time value of money* – for use when making comparisons between alternatives exhibiting differential timing in the payment and receipt of cash.

Economic life
The period of time during which investment in an asset is the least-cost alternative for satisfying a particular objective.

Element
The portion of a project that fulfils a particular physical purpose irrespective of construction and/or specification.

Elemental cost analysis
The technique of *cost control* involving the division of the project into element and sub-element groupings to enhance control and facilitate comparison with similar elements and sub-elements in other buildings.

Escalation
An increase in the price of a particular good or service. Sometimes distinguished from *inflation*, but often the terms are used interchangeably.

Future value
The comparative value of costs and benefits to future generations determined after the adjustment of *present value* for changes in worth.

Inflation
A continuing increase in general price levels. Sometimes distinguished from *escalation*, but often the terms are used interchangeably.

Internal rate of return
The discount rate that leads to a *net present value* of zero.

ISO
The International Organization for Standardization is an international non-governmental organisation that promotes the

development and implementation of voluntary international standards.

Maintenance
Combination of all technical and associated administrative actions during the service life to retain a facility or its parts in a state in which it can perform its required functions.

Net present value
The sum of the discounted present (or future) value of all expected cash inflows and outflows for a project over a selected *time horizon*.

Obsolescence
Loss of ability of an item to perform satisfactorily due to changes in performance requirements.

Occupancy cost
The costs of staffing, manufacturing, managing supplies and the like that relate to the building's function.

Operating cost
The expenditure required to maintain the land and building and facilitate its function.

Ownership cost
Regular running costs such as cleaning, rates, electricity and gas charges, insurance, maintenance staffing, security, and the like.

Performance evaluation
Evaluation of critical properties on the basis of measurement and inspection.

Present value
The real value of costs and benefits to the present generation equal to the future cost expressed in constant money – used for measurement and control purposes and as the basis of the whole life plan.

Repair
Return of a facility or its parts to an acceptable condition by the renewal, replacement or mending of worn, damaged or degraded parts.

Residual life
The remaining component life at the end of the selected *time horizon*.

Risk
The measure of the level of uncertainty of future events.

Risk analysis
The deterministic or probabilistic techniques for the identification and assessment of *risk*.

Service life
The period of time after installation during which a facility or its parts meets or exceeds the performance requirements.

Time horizon
The period of time covered by an economic appraisal or whole life plan and recommended to be determined by the period of financial interest of the owner or investor.

Time value of money
The concept that money changes over time, as a result of opportunity or time preference considerations.

Whole life
The period of time between the baseline date and the time horizon, over which the future costs relating to the decision or alternative under study will be incurred.

Whole life analysis
That part of the cost control process dealing with the monitoring and recording of whole life costs, the compilation of feedback and the management of future whole life cost performance during construction and occupation.

Whole life cost
Total cost of an asset measured over a period of financial interest of the owner (often called life cycle cost and cost-in-use).

Whole life planning
That part of the cost control process dealing with budgeting, optimisation and documentation of whole life costs during design.

References and bibliography

Abdou, A. (2000) *Collaborative Management Tools in the Information Technology Era*, PMI-AGC 7th Annual Conference (Arabian Gulf Chapter), 7 February, Bahrain.

Al-Hajj, A., Pollock, R., Kishk, M., Aouad, G., Sun, M. and Bakis, N. (2001) *On the requirements for effective whole life costing in an integrated environment*, COBRA conference papers, Glasgow Caledonian University, RICS Foundation.

Aouad, G., Bakis, N., Amaratunga, D., Osbaldiston, S., Sun, M., Kishk, M., Al-Hajj, A. and Pollock, R. (2001) *An integrated life cycle costing database – a conceptual framework*, ARCOM conference, The University of Salford.

Armitage, A. (2002) The black hole, *Building*, 22 March, pp. 74–75.

Blanchard, B.S. (1992) *Logistics and Engineering Management*, 4th edition, Prentice Hall, New Jersey.

BMI (2001) *Life expectancy of building components: surveyors' experiences of building in use – a practical guide*. Building Market Information, The Royal Institution of Chartered Surveyors, UK.

BMI (2003) *Review of occupancy costs 2003*, BMI Special report Serial 322, Building Market Information, The Royal Institution of Chartered Surveyors, UK.

Bourke, K. (2001) Whole life costing – an overview, in Construction Industry Research and Information Association, *Service life planning/whole life costing*, Report of a joint workshop with the RICS and RIAS held at the Scottish Executive, Victoria Quay, Edinburgh, 7 February, Members' Report E1102.

Boyce, D. (2002) Logistics with flying colours, *Project*, June, Volume 15, Issue 2, pp. 10–11.

Brown, S. (2002) Raising a glass – by using technology, *Project*, June, Volume 15, Issue 2, pp. 22–23.

BSI (1992) British Standard 7543:1992 – Guide to durability of buildings and building elements, components and materials. British Standards Institute, London.

Central Data Management Authority (2001) *Making information make sense*, Ministry of Defence, London.

Clift, M.R. and Butler, A. (1995) *The performance and costs-in-use of buildings: a new approach*, Building Research Establishment Report BR 277, Building Research Establishment, Watford.

Clift, M. and Bourke, K. (1999) *Study on whole life costing*, Building Research Establishment Report Number CR 366/98, Building Research Establishment, Watford.

Construction Best Practice Programme (2001) Themes – Whole Life Costing. www.cbpp.org.uk/

Construction Industry Research and Information Association (2001a) Asset management, *CIRIA News*, Issue 1, 6.

Construction Industry Research and Information Association (2001b) *Service life planning/whole life costing*, Report of a joint workshop with the RICS and RIAS held at the Scottish Executive, Victoria Quay, Edinburgh, 7 February, Members' Report E1102.

CRISP (1999) UK Study on Whole Life Costing. Construction Research and Innovation Strategy Panel Performance Theme Group, http://www.crisp-uk.org.uk/

Dell'Isola, A.D. and Kirk, S.D. (1981) *Life Cycle Costing for Design Professionals*. McGraw Hill Book Company, USA, p. 224.

Department of Defense (1983) *Logistic Support*, Military Standard 1388-1A, USA.

Department of the Environment, Transport and the Regions (2000) *Asset management of Local Authority land and buildings – good practice guidelines*, DETR, London.

DETR (1998) Construction Task Force Report, *Rethinking Construction*, DETR, London.

DTI (2001) *The State of the Construction Industry Report*, February 2002.

Dunn, R. (1987) Advanced maintenance technologies, *Plant Engineering*, Vol. 40, pp. 80–2.

Edwards, S. Bartlett, E, and Dickie, I. (2000) *Whole life costing and life cycle assessment for sustainable building design*, Digest 452, BRE, Watford.

Egan, Sir John (2002) *Accelerating Change – a consultation paper by the Strategic Forum for Construction*, DETR, London.

El-Haram, M. and Horner, M.W. (2002) Practical application of RCM to local authority housing: a pilot study, *Journal of Quality in Maintenance Engineering*, Vol. 8, No. 2, pp. 135–143.

El-Haram, M., Marenjak, S. and Horner, M. (2002a) *The use of ILS Techniques in the Construction Industry*, The University of Dundee.

El-Haram, M., Marenjak, S. and Horner, R.M.W. (2002b) Development of a generic framework for collecting whole life cost data for the building industry. *Journal of Quality in Maintenance Engineering*, **8**, 2 July 2002, 152–160.

EC Harris (2002) Global Review 2nd quarter 2002, EC Harris, London, UK, p. 1.

Fairs, M. (2002a) The logistics revolution, *Building*, 28 June, Issue 25, pp. 40–48.

Fairs, M. (2002b) What the talking toilet has to tell us, *Building*, 14 June, Issue 23, pp. 22–23.

Finch, E. (2000) *Net Gain in Construction. Using the Internet in the Construction Industry*, Butterworth Heinemann, Oxford.

Finkelstein, W. (1988) *Integrated logistics support: the design engineering link*, IFS: Bedford.

Flanagan, R., Ingram, I. and Marsh, L. (1998) *A bridge to the future – profitable construction for tomorrow's industry and its customers*, Thomas Telford Publishing, London.

Flanagan, R. and Norman, G. (with Furbur, J.D.) (1983) *Life Cycle Costing for Construction*, Surveyors Publications, London.

Gordon, A.R. and Shore, K.R. (1998) *Life Cycle Renewal as a Business Process*, Innovations in Urban Infrastructure Seminar of the APWA International Public Works Congress, Las Vegas, USA, pp. 41–53, available at: http://www.nrc.ca/irc/uir/apwa

Government Construction Procurement Group (1999) *Financial Aspects of Projects*, Procurement Guidance No. 6, HM Treasury, London.

HMSO (1995) *Progress through Partnership*, Office of Science and Technology, HMSO, London.

Hoar, D. and Norman, G. (1990) Life cycle cost management. In Brandon, P. (ed.) *Quantity Surveying Techniques: New Directions*, BSP Professional Books, UK.

ISO 15686-1 (2000) *Building and constructed assets – service life planning*, International Standard, Part 1: General principles, BSI, London.

Jackson, S. (2002) *Development of a whole life design appraisal tool*, Proceedings of 18th Annual Conference of the Association of Researchers in Construction Management, University of Northumbria, 2–4 September.

John, G., Loy, H., Clements-Croome, D., Fairey, V. and Neal, K. (2002) *Enhancing the design of building services for whole life performance*, The University of Reading.

Kamasho, C. (2002) Turning information into knowledge, *Chartered Surveyor Monthly*, March 28–29.

Lane, T. (2002) Palm stormers, *Building*, 8 February, 58–59.

Mackay, A.D. (2001) *Setting the Scene*, Keynote Speech, The whole-life performance of facades, Proceedings of a national conference organised by the Centre for Window & Cladding Technology, Ledbetter, S. and Keiller, A. (eds), 18/19 April, The University of Bath, pp. 13–20.

Madine, V. (2002) Whatever happened to best value?, *Building*, 19 April, pp. 48–49.

Ministry of Defence (1998a) *Logistic Support Analysis (LSA) and Logistic Support Analysis Record (asset register)*, Defence Standard 00-60 Part 1, Issue 2, 31 March, MOD, Glasgow.

Ministry of Defence (1998b) *Guidance for Application of Software Support*, Defence Standard 00-60 Part 3, Issue 2, 31 March, MOD, Glasgow.

Ministry of Defence (1998c) *Procedures for Procurement Planning*, Defence Standard 00-60 Part 23, Issue 2, 31 March, MOD, Glasgow.

Ministry of Defence (1998d) *Procedures for Invoicing*, Defence Standard 00-60 Part 25, Issue 2, 31 March, MOD, Glasgow.

Ministry of Defence (2002) *MOD Guide to Integrated Logistics Support*, Issue 2, MOD, Bath.

Ministry of Defence (2002a) *Application of Integrated Logistic Support (ILS)*, Defence Standard 00-60 Part 0, Issue 5, 24 May, MOD, Glasgow.

Ministry of Defence (2002b) *Guide to the application of LSA and asset register*, Defence Standard 00-60 Part 2, Issue 5, 24 May, MOD, Glasgow.

Ministry of Defence (2002c) *Electronic documentation*, Defence Standard 00-60 Part 10, Issue 5, 24 May, MOD, Glasgow.

Ministry of Defence (2002d) *Application of Integrated Supply Support Procedures*, Defence Standard 00-60 Part 20, Issue 6, 24 May, MOD, Glasgow.

Ministry of Defence (2002e) *Procedures for Initial Provisioning*, Defence Standard 00-60 Part 21, Issue 5, 24 May, MOD, Glasgow.

Ministry of Defence (2002f) *Procedures for Codification*, Defence Standard 00-60 Part 22, Issue 3, 31 January, MOD, Glasgow.

Ministry of Defence (2002g) *Procedures for Order Administration*, Defence Standard 00-60 Part 24, Issue 5, 24 May, MOD, Glasgow.

Ministry of Defence (2002h) *Procedures for Repair Administration*, Defence Standard 00-60 Part 26, Issue 1, 8 February, MOD, Glasgow.

Moubray, J. (1997) *Reliability-centred Maintenance*, Butterworth-Heinemann, Oxford.

National Audit Office (2002a) *Ministry of Defence – Major repair and Overhaul of Land Equipment*, Report by the Comptroller and Auditor General, HC 757, Session 2001–2002, 26 April, NAO, London.

National Audit Office (2002b) *Ministry of Defence – Helicopter Logistics*, Report by the Comptroller and Auditor General, HC 840, Session 2001–2002, 23 May, NAO, London.

National Audit Office (2002c) *Ministry of Defence – Progress in Reducing Stocks*, Report by the Comptroller and Auditor General, HC 898, Session 2001–2002, 20 June, NAO, London.

National Audit Office (2002d) *Ministry of Defence – Exercise Saif Sareea II*, Report by the Comptroller and Auditor General, HC 1097, Session 2001–2002, 1 August, NAO, London.

OGC (2002) *Construction Procurement Guide No. 7 Whole Life Costs*, Office of Government Commerce, London, UK.

Paz, N.M. and Leigh, W. (1994) Maintenance scheduling: issues, results and research needs, *International Journal of Operations and Production Management*, Vol. 14 No. 8, pp. 47–69.

Property Advisers to the Civil Estate and Office of Government Commerce (2000) *Achieving Excellence – The Government Construction Clients' Panel, GC/Works/1 Amendment 1 (2000)*, The Stationery Office, Norwich.

Ribeiro, F.L. (2001) *Internet and web technology for knowledge management in construction organisations*, COBRA conference papers, Glasgow Caledonian University, RICS Foundation.

Ribeiro, F.L. and Lopes, J. (2001) *Knowledge-based e-business in the construction supply chain*, COBRA conference papers, Glasgow Caledonian University, RICS Foundation.

RIBA (2000) *The Architect's Plan of Work*, RIBA Publications, London.

RICS (1999) Part Two: Construction Design and Economics. Section 2: Life Cycle Costing. The Surveyor's Construction Handbook (effective from 1999), RICS, London, UK.

Royal Academy of Engineering (1998) The Long Term Cost of Owning and Using Buildings, raeng.org.uk/

Simm, J. and Masters, N. (2003) Whole life costs and project procurement in port, coastal, and fluvial engineering. HR Wallingford.

Smith, P. (1999) Occupancy cost analysis, in Best, R. and Valence, G. de (eds) *Building in Value: Pre-design Issues*, Edward Arnold, London.

Sustainability Action Group of the Government Construction Clients' Panel (2000) *Achieving sustainability in construction procurement – sustainability action plan*, The Stationery Office, Norwich.

Watson, C. (1999) A building is for life, not just for sale, *ASI Journal*, p. 14.

Wornell, P. (2000) Construction Audit Ltd: whole life cost – the potential for insuring the risks, in Construction Industry Research and Information Association (2000) *Whole life costing*, report of a workshop held at 100 Temple Street, Bristol, 7 December, Members' Report E0134.

US Department of Energy (1995) *Life Cycle Costing Manual for the Federal Energy Management Program*, US Department of Commerce.

Index